THERMAL MANAGEMENT OF TELECOMMUNICATIONS EQUIPMENT

THERMAL MANAGEMENT OF TELECOMMUNICATIONS EQUIPMENT

L. T. Yeh
R. C. Chu

ASME PRESS

ASME Press Book Series on Electronic Packaging
Series Editor Dereje Agonafer

Library of Congress Cataloging-in-Publication Data

Yeh, L.-T. (Lian-Tuu), 1944-
Thermal management of telecommunications equipment / L.T. Yeh, Ph.D., P.E., R.C. Chu. pages cm
Includes bibliographical references and index.
ISBN 978-0-7918-6020-5 (alk. paper)
1. Electronic apparatus and appliances--Thermal properties. 2. Telecommunication--Equipment and supplies--Cooling. 3. Telecommunication--Equipment and supplies--Protection. 4. Heat--Transmission. I. Chu, R. C. (Richard C.), 1933- II. Title.
TK7870.25.Y443 2013
621.382--dc23

2013027405

TABLE OF CONTENTS

LIST OF FIGURES

FOREWORD

Miniaturization and convergence are key drivers in microelectronics and telecommunications in general. Cooling of telecommunication systems is quite a challenge as the scales vary from hand held devices to large systems. This is a comprehensive book that addresses both the thermal and mechanical reliabilities in the design of telecommunication systems. It is important to note that both authors have a significant industrial experience and as such, the guidance provided in this book stems from a significant practical experience.

It is very sad that recently, Dr. Richard C. Chu has passed away. Dr. Chu has been recognized by many as the "father of electronic cooling." Dr. Chu's invention achievements are documented in over 200 patents and over 150 patent disclosure publications. A prolific technical innovator in cooling technology, many of his patents and publications have been recognized outside of IBM as industry standards for the cooling of electronic equipment. Dr. Chu has achieved numerous honors, including the ASME Heat Transfer Memorial Award in 1986, election to the National Academy of Engineering in 1987 and his appointment to Academia Sinica, Republic of China in 1996. He served as the first president of the IBM Academy of Technology and was recognized by the Organization of Chinese Americans (OCA) as Asian American Engineer of the Year in 2006. In 2010, Dr. Chu received his 66th Invention Achievement Award, at that time a record for IBM Poughkeepsie. Dr. Chu had a lifelong passion for the academic community, as shown by his active involvement in IBM's university research program which resulted in over 80 thesis-related publications. Dick retired in 2011 after 51 years of dedicated service to IBM. I worked for Dick Chu for 15 years at IBM and consider him my mentor. He is by far more than anyone else responsible for my career path guiding me during every fork in the road. Likewise, Dr. Yeh considers Dick his mentor and dedicates this work to Dick's memory.

The book can be used as an elective senior level or graduate level textbook. It is also a must for engineers in the field who are always led to finding best practices with the ever shrinking of lead time for designing systems.

Dereje Agonafer, ASME & AAAS Fellow
Jenkins Garrett, Professor
Site Director, NSF IUCRC Center in Energy Efficient Systems
Director, Electronics, MEMS & Nanoelectronics Systems Packaging Center
University of Texas at Arlington
Editor in Chief, ASME Book Series on Electronic Packaging

PREFACE

The need to develop reliable microelectronic devices capable of operating at high speeds with complex functionality requires a better understanding of the factors that govern the thermal performance of electronics. With an increased demand on system reliability and performance combined with miniaturization of the devices, thermal consideration has become a crucial factor in the design of electronic packages, from chip to system levels.

The authors understand that the challenge in the field of thermal management of an electronic system resides not only with very high performance and high heat dissipation devices, but also with the intermediate and lower power devices where improved reliability objectives require cooler operation of chips. The authors further realize that no one design method is best suited for all applications. It is common to employ several different heat transfer modes simultaneously in a system. Therefore, the authors published their first book entitled "Thermal Management of Microelectronic Equipment" in 2002 to address a wide range of subjects related to various heat transfer technologies that can be utilized for thermal design of electronic equipment.

In the recent years, telecommunication product development has gone through tremendous changes due to demands of high internet speeds and capacities as well as system reliability. These changes result in a significant increase in system power requirements which makes thermal design much more difficult and challenging. Furthermore, equipment power consumption is continuously increased at a fast pace. Since the current telecommunication systems are still focusing air cooling, large high-performance fans are becoming a must for high power systems in order to provide sufficient air flow rates. In addition to system thermal issues, two major concerns about these large fans are power consumption and acoustic noise of the fans. The increased system power results in a significant increase in power consumption and the cost of powering and cooling of equipment as well as the data centers in the last few years. Therefore, energy efficiency at the telecommunication equipment level and in central office/data center becomes the top priority of equipment manufacturers and operators of the facilities.

This book covers all types of telecommunication equipment, including indoor and outdoor systems. There are ten chapters in the book. Among them, Chapters 2–4 focusing on the understanding of the physics of individual components (or parts) such as chip package, heat sink, fan and air handling system.

Chapters 5–7 provide basic methodology for system thermal design on any indoor system. The equipment under consideration includes large telecom racks or cabinets, and small box configurations. Both active and passive cooling is employed, depending on system configuration and power. The detailed descriptions of the system thermal design for any outdoor equipment are included in Chapter 8. The outdoor systems are generally classified into two groups, the ground/vehicle mounted and tower/pole mounted equipment. All outdoor equipments are subjected to extreme weather conditions, high humidity, high ambient temperature and solar radiation. At the same time, they are expected to operate with high reliability. A large variety of cooling techniques are available in cooling of the outdoor electronics as well as telecommunications equipment. These include passive and active forced air cooling, the air conditioners or heat pumps, thermosyphons and phase change materials (PCMs). To pave the way for future high power telecom systems, the available liquid cooling technologies are presented in Chapter 9. Finally, Chapter 10 serves as a summary of the entire book which includes the listing the future research needs. An extensive discussion of the energy efficiency at the equipment and data centers, including the utilization of alternative energy is also included in this chapter.

This book places a great deal of emphasis on providing practical solutions to the thermal issues facing in today's telecommunication industry. Therefore, all materials presented in the book are based on authors' experience in design of all types of the telecommunication equipment. This book is developed to serve many types of readers. For graduate students, this book will guide them by practical approaches to solve real world problems that are of a vast complexity. For professional engineers, this book will provide them with valuable and resourceful references.

L. T. Yeh
R. C. Chu

Chapter 1

INTRODUCTION

Telecommunication, which is defined as the use of information transmission over some significant distances to communicate, employs the electrical devices and the use of radio and microwave communications as well as the fiber optics and their associated electronics plus the use of the orbiting satellites and the Internet. A telecommunication network is a collection of terminals, links, and nodes which connect together so that communications can be made between users of the terminals. Networks include the circuit switching or message switching. The terminals in the network must have a unique address so the messages or connections of addresses in the network can be routed to the proper terminal/recipients. The telecommunication network generally consists of several basic components which include terminals, telecommunications processors, telecommunications channels, computers, and telecommunications control software.

The telecommunication equipment can be deplored in the indoor and the outdoor environments depending on its applications. Based on the functionality and performance, the indoor systems include, but are not limited to, the switch/router, the optical network, the metro-ethernet, the access, and the core network. A brief description of them is as follows:

Router: The network router is a device that forwards and routes data packet along networks. Two basic functions of a router are: (a) to insure information does not go where it is not needed and (b) to make sure information to the intended destination. A router is often included as part of a network switch.

Switch: The switch is a device that channels incoming data from any of multiple input ports to the specific output port that will take the data toward its intended destination. There are two types of switching methods, circuit-switching and packet-switching. The former is really a dedicated and continuously connected path for its duration. In other words, a network's path can be used exclusively for certain duration for tow or more parties and then switched for use to another set of parties. For example, an ordinary voice phone call generally is utilizing a circuit-switching. Most of today data using digital signal is sent over network that use packet-switching which allows users to share the same paths at the same time. In addition, the specific route a data unit travels can also be varied as conditions change, while access services are specified sets of information transfer capabilities furnished to users at telecommunications network points-of-termination (POTs) to provide access to network transport services.

A core network is the central part of a telecommunication network that provides various services to customers who are connected by the access network. All of these systems have the same basic mechanical configuration. For the mechanical configurations, the equipment is generally classified into the rack/cabinet and box configurations as shown in Figure 1.1.

The explored growth of the internet usage in recent years has made the telecommunication product development go through tremendous changes due to demands of high internet speeds and capacities. These changes result in a significant increase in the system power requirements as did in microelectronic systems. Therefore, thermal management becomes most critical factor in the design of any telecommunication equipment.

Though liquid cooling has long been adopted in the computer industry, air cooling is still the main cooling scheme in the telecommunication industry. The key components in any telecommunication system are the microelectronics, the printed circuit boards (PCBs), and the fans. Other relevant standards are listed in the References.

The outdoor systems include the outside plant (OSP) and the remote radio head (RRH). The former includes both landline and the wireless applications, while the latter is generally for the wireless applications. Both systems are illustrated in Figure 1.2. The RRH is a pole or tower mounted equipment. Therefore, the size and the weight are the critical factors to be considered in the design of the RRH.

The design and operation of the telecommunication equipment is mainly governed by GR-63 [1] and GR-487 [2] for the indoor and outdoor systems, respectively.

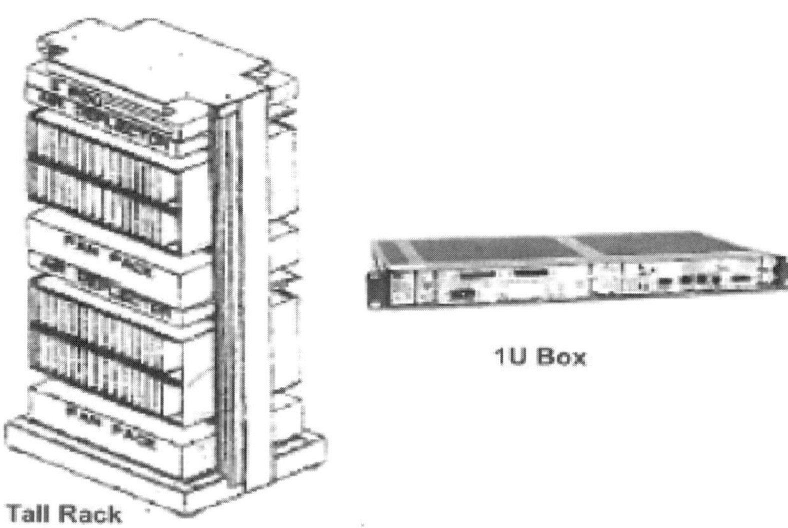

FIGURE 1.1 Typical telecommunication cabinet and box configuration.

OSP Cabinet　　Remote Radio Head

FIGURE 1.2 Typical outside plant and remote radio head.

REFERENCES

1. Telcordia Technologies Generic Requirements, GR-63-CORE, Issue 3, March, 2006 (NEBS Requirements: Physical Protection).
2. Telcordia Technologies Generic Requirement, GR-487-CORE, Issue 3, April, 2009 (Generic Requirements for Electronics Equipment Cabinets).
3. Telcordia Technologies Generic Requirement, GR-3028-CORE, Issue 1, December, 2001 (Thermal Management in Telecommunications Central Offices).
4. Telcordia Technologies Generic Requirement, GR-950-CORE, Issue 2, December, 1998 (Generic Requirements for Optical Network Unit (ONU)).
5. Telcordia Technologies Generic Requirement, GR-3108-CORE, Issue 1, July, 2004 (General Requirements for Network Equipment in the Outside Plant (OSP)).

Chapter 2

COMPONENT AND PRINTED CIRCUIT BOARD

2.1 COMPONENT

Since the invention of an integrated circuit (IC) in the latter part of the 1950s, the progress of development of new integrated circuits has been advanced very rapidly. Generally, heat is generated from the transistors on a small, thin, and fragile silicon die or chip inside the package. The chip package that is the housing for the silicon die serves to protect the chip from the environment and to facilitate the handling during the manufacturing process as well as chip interconnection.

2.1.1 Chip Packaging Technology

The most common methods of chip level interconnects are wire bonding, tape automated bonding (TAB), and solder bumping. Among them, the solder bumping flip chip provides the highest packaging density with less packaging delay. The main discussion here is to focus at the flip chip package because of its explosive growth in recent years. The flip chip packaging, which is also known as Controlled Collapse Chip Connection (C4), is a method for interconnecting semiconductor devices such as IC chips to external circuitry with solder bumps. The flip chip packaging has several advantages over the wire bond packaging and these are as follows:

1. Higher I/O density
2. Better electric performance (i.e., for high speed or high frequency design)
3. Smaller package and lower cost
4. Improved thermal management

Figure 2.1 shows a typical flip chip package with a heat sink. The flip chip is a chip packaging technique in which the active area of the chip is flipped over facing downward. The most common bumping technologies are that the metalized (solder) pads are deposited on the top side of the wafer. The chip is flipped over

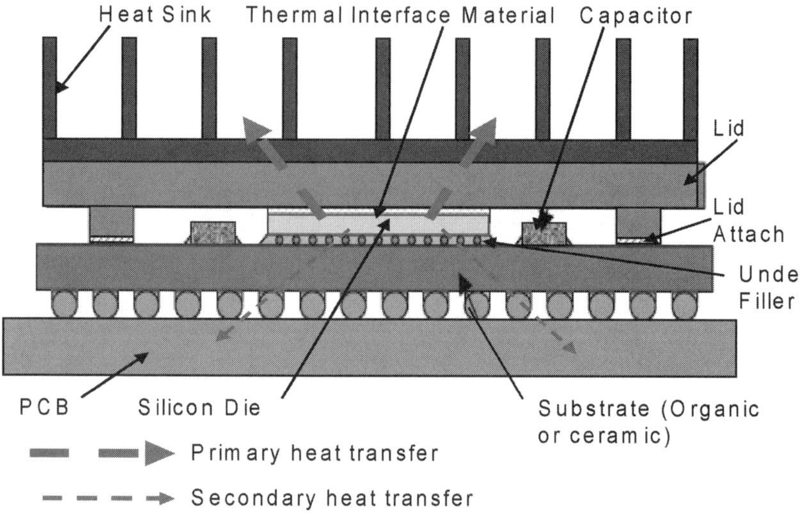

FIGURE 2.1 Typical flip chip package with heat sink.

with its top side facing down and is soldered to the substrate are as illustrated in Figure 2.2 [1]. The solder ball and column are made of high melting point (> 300°C) solders such as 95/5 and 90/10 lead/tin compositions, which do not reflow during package or board assembly.

An epoxy is employed to underfill the gap between the flip chip and the substrate. The filling epoxy completely covers the flip chip interreconnections and makes a strong bond to both the device and the substrate. The underfill is needed because the shear stress generated by the CTE (coefficient of thermal expansion) mismatch between the chip and the substrate will lead to the solder joint failure due to the fatigue. The underfill enhances the fatigue life of the solder interconnection by as much as 10 times and also provides the protection from the environments.

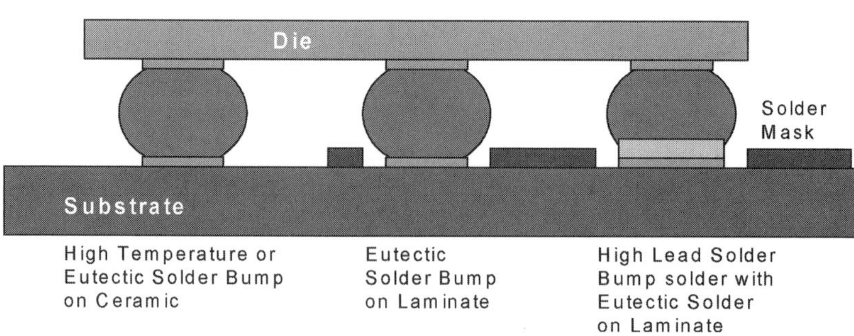

FIGURE 2.2 Common solder bumping technologies.

2.1.2 Thermal Management

As indicated in Figure 2.1, the majority of heat generated from the chip flows to the lid (or case) and then to the heat sink and finally heat is transferred to the air stream over the heat sink. On the other hand, the minority of heat is conducted from the chip to the printed circuit board, and through the spreading over the board, heat is finally transferred to the air stream over the board.

The thermal representation of the flip chip package as illustrated in Figure 2.1 is given in Figure 2.3. The thermal interface material (TIM) as shown in Figure 2.1 is directly applied on the back side of the silicon die which provides a good thermal path to the lid, and then to the heat sink, and finally to the ambient air stream. The material properties of a TIM must be low thermal resistance, low stress, and good adhesion. As chip power increases, the role of the TIM for the flip chip package becomes increasingly important.

Reference 2 performs CFD analysis to identify the key design parameters for TIM in different applications. The following design guidelines are recommended based on the analysis results:

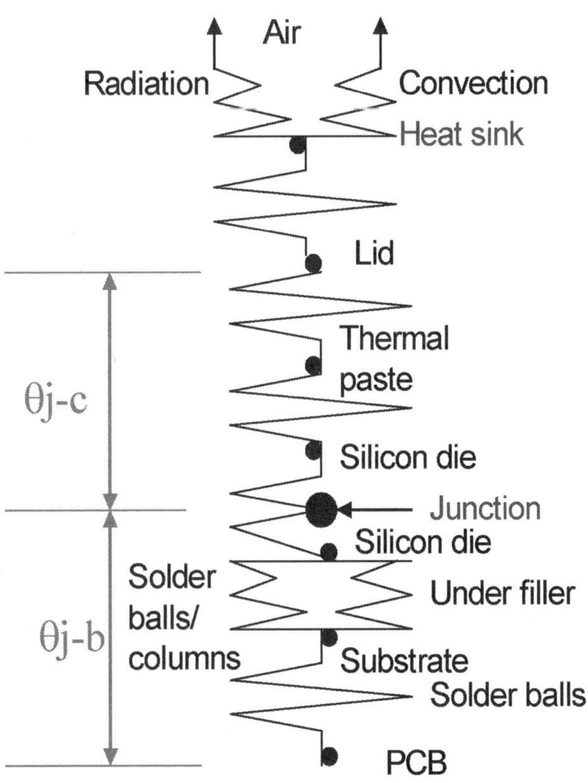

FIGURE 2.3 Thermal representation of flip chip and heat sink.

1. For chip power less than 10 W, the selection of TIM is less stringent because thermal performance of packages is not very sensitive to the changes to thermal conductivity, area coverage and thickness of TIM as evidenced below
   TIM thickness: ≤ 500 μm
   TIM coverage: ≥ 25% of die area
   TIM thermal conductivity: ≥ 1 W/m °C after degradation
2. For chip power ranging between 10 and 30 W, where a heat sink is required, TIM must be carefully selected as described below
   TIM thickness: ≤ 200 μm
   TIM coverage: ≥ 50% of die area
   TIM thermal conductivity: ≥ 2 W/m °C after degradation
3. For chip power ranging between 30 and 100 W, TIM selection becomes critical and must meet the following minimum requirements
   TIM thickness: ≤ 150 μm
   TIM coverage: ≥ 75% of die area (recommendation of 100% coverage for power greater than 50 W)
   TIM thermal conductivity: ≥ 5 W/m °C after degradation

Because of the short thermal path, the primary heat transfer path from the flip chip is through the TIM on the back side of the die to the lid as shown in Figure 2.1. The total thermal resistance from the die to the lid is referred to as the thermal resistance from the junction to case (**θj-c**). On the other hand, the total thermal resistance from the die to the board is referred to as the thermal resistance from the junction to board (**θj-b**) which is generally several times larger than **θj-c**. This is the reason that the heat transfer from the die to the board is the secondary heat transfer path. The heat transfer rate from the die to the case in the primary path is typically greater than 75% of the chip power. The more effective the heat sink, the higher the heat transfer rate through the primary heat transfer path is.

The lid is a critical part of the package. The lid provides the structural support to the heat sink and also protects the package. In addition, the lid basically acts as a heat spreader to reduce the non-uniform temperature distribution over the lid which can lower the temperature gradient across the interface between the package and the heat sink. Lids can be made from many high thermal conductivity materials such as copper, aluminum, AlSiC, AIN, CuW, and even alumina. However, it is critical for the lid to have matching CTE as closer to Silicon die or Gallium Arsenide (GaAs) chip as possible. Figure 2.4 shows the thermal conductivity and CTE of various materials [3].

Ramakrishna and Lee [4] conducted a detailed study to examine the effects of various parameters on thermal performance of a flip-chip-plastic ball grid array (FC-PBGA) package which is illustrated in Figure 2.5. The figure shows the cross-section view of the package in a wind tunnel in the simulations. A schematic diagram of various enhancements under consideration is given in Figure 2.6. Though the figure includes all enhancements, however, all of them are studied individually and separately.

Traditional thermal performance metrics used to specify package thermal performance are the thermal resistance of junction-to-air (**θj-a**), thermal resistance of junction-to-board (**θj-b**), and thermal resistance of junction-to-case (**θj-c**) which are defined below

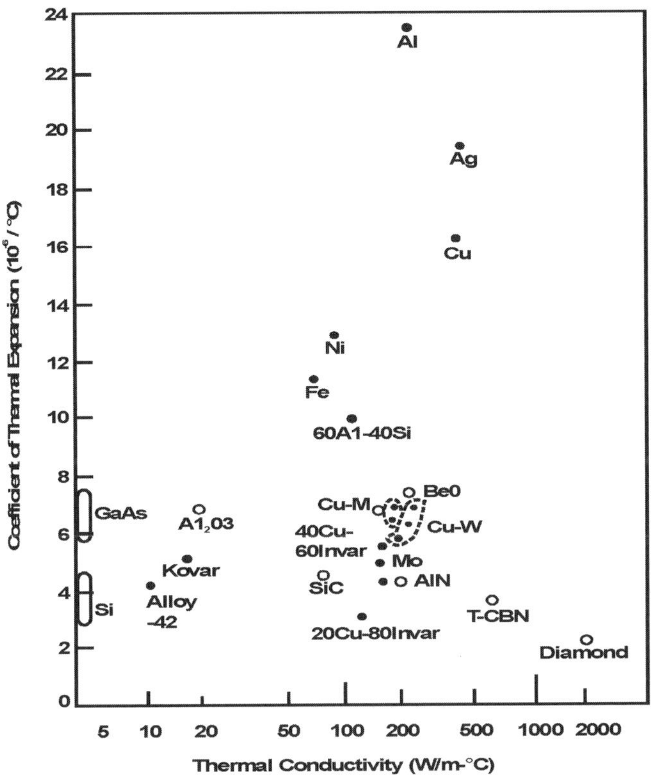

FIGURE 2.4 Thermal conductivity and CTE of various materials.

$$\theta j - a = (Tj, max - Ta)/P \tag{2.1}$$

$$\theta j - b = (Tj, max - Tpcb)/P \tag{2.2}$$

$$\theta j - c = (Tj, max - Tc)/P \tag{2.3}$$

where Ta, Tc, Tj, and Tpcb are the temperature of inlet air, component case, component junction, and board, respectively, and P is component (package) power.

The parameters under consideration in the analysis are as follows:

a. Effect of C4 and underfill thermal conductivity
b. Effect of thermal balls and PTHs in PCB
c. Effect of heat spreader with and without overmolding

The analysis was performed under both natural convection and forced convection. The parametric study includes die area of 48.4 and 118.2 mm², substrate

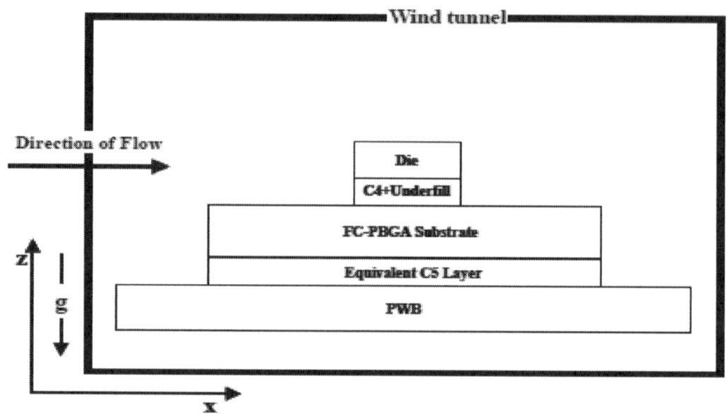

Figure 2.5 Cross-section of package in wind tunnel.

sizes of 25 to 35 mm and the free stream velocity 0.5 to 2 m/sec for forced convection air cooling. The goal of the analysis is to obtain the package thermal resistances as described in Equations (2.1) to (2.3) which are then used to evaluate the effects of various parameters on package thermal performance.

Based on this study, the following conclusions can be made:

a. Effect of C4 and underfill thermal conductivity
   – The effect of combined thermal conductivity of the C4 and underfiller materials in range between 0.6 and 10 W/m °C on package (FC-PBGA) thermal performance, (e.g., θj-a, θj-b, and θj-c) is negligible for its value greater than 3 W/m °C and some limiting improvements for those with

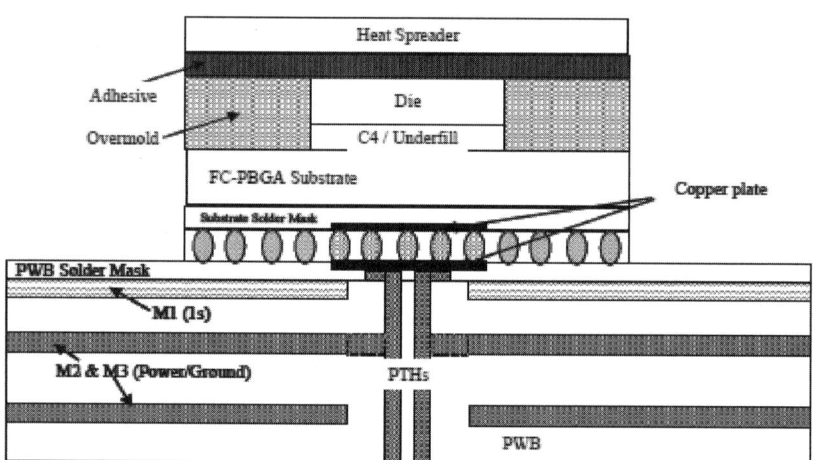

Figure 2.6 Schematic diagram of various thermal enhancements.

thermal conductivity smaller than 3 W/m °C. 0.6 to 10 W/m °C is the underfill thermal conductivity.
   – The thermal performance is significantly improved (decreasing in package thermal resistance) with increasing the die size or air velocity.

b. Effect of thermal balls and PTHs in PCB
   – The C5 (controlled collapse chip connection carrier) connections between the package and the board provide a low heat flow path. However, thermal balls (C5) connected to the PTHs in the printed circuit board can decrease package thermal resistance by about 10% to 15% as the number of thermal balls and PTHs increase zero to 9 × 9 on 1.27 mm pitch. The effect of the die size on this thermal enhancement is more profound on the junction to board thermal resistance (θj-b) than on the junction to air resistance (θj-a).

c. Effect of heat spreader with and without overmolding
   – The heat spreader on the back of the die decreases θj-a by about 7% in natural convection and by 28% in forced convection with air velocity at 2 m/sec for the die size of 48.4 mm^2. The enhancement provided by a heat spreader (without overmold) in natural convection for large die sizes is so small that its use is not justified in view of the added cost of materials and assembly.
   – The overmolded die with a heat spreader on the top of the overmold provides better thermal enhancement than a heat spreader on the die for air velocities up to 1 m/sec. The heat spreader without overmold performs slightly better than that with heat spreader plus overmold for the air velocity greater than 1 m/sec.

The silicon die generally includes multiple heat sources. The size and power of each heat source are different which results in the local hot spots. The shape of the heat source on the silicon die actually has a great effect on the junction temperatures of the chip. However, it seems that no study has been made to address this issue.

The sketch as illustrated in Figure 2.7 will be used as an example for discussion. The figure includes three heat sources on a single die. All of them have the same area but with different length of the perimeter. Heat generated at the source area is first conducted through the die thickness, then the TIM, and finally to the lid. Along these heat conduction paths, heat is also spread out laterally within each layer. Therefore, the larger the perimeter, the better the heat spreading will be because of the larger lateral area for heat conduction.

2.1.3 Advanced Packaging Technology

Due to the increased power requirement, the vertical integration of the circuit die which is also referred to as "3D Packaging" has received considerable attention in the recent years. One of the examples of 3D package is proposed by IBM [5] as shown in Figure 2.8. Thermally, the vertically stack chip packages are even more demanding than the traditional 2D chips. Heat is generally conducted through

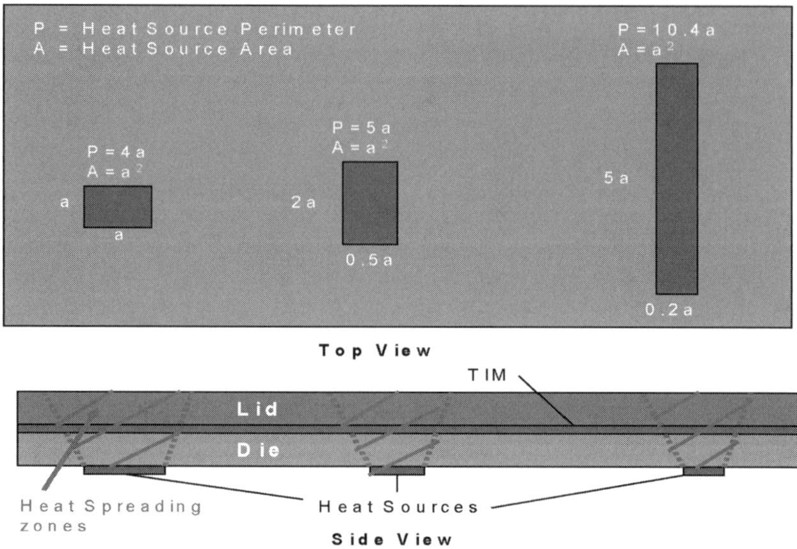

FIGURE 2.7 Heat conduction spreading in die.

the multiple interfaces and layers of the different materials to the case (lid) and to the board in the traditional 2D package. The total thermal resistance is too large to meet the temperature limits for the interlayer of the chip package due to limiting thermal paths from the interlayers to the ambient.

The major challenge in the thermal design of the 3D chip package is how to remove heat from interior layers of the dies. For these high power 3D chips, the traditional air cooling techniques are not possible because of extremely poor thermal properties of the air. It will require unrealistically high airflow rates for effective air cooling. The microchannel water cooling on the integrated circuits was

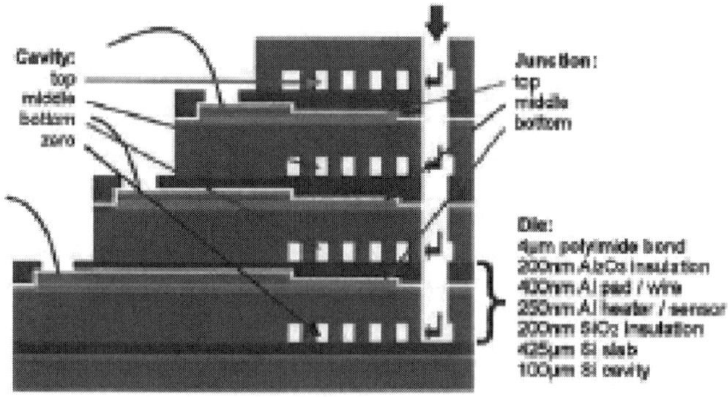

FIGURE 2.8 IBM 3D packages.

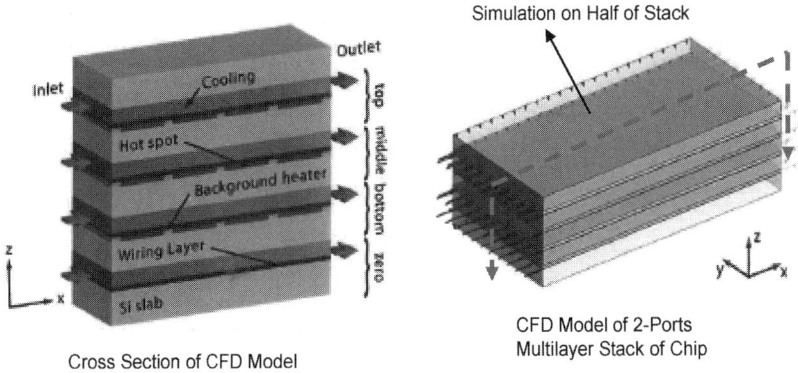

FIGURE 2.9 Cross-section of CFD thermal model.

first proposed by Tuckerman and Pease [6]. The backside heat removal using the microchannel cooling technology has been shown effective for the single layer die. However, it will not provide adequate cooling for the interlayer dies. The interlayer integrated water cooling as shown in Figure 2.8 is a very promising approach.

As can be seen in Figure 2.8, the thermal design is based on the interlayer integrated cooling concept. The cooling network, which includes micro pin-fins and microchannel heat transfer structures, is integrated between the layers. The authors not only perform experimental work but also develop the porous medium model to simplify thermal model as illustrated in Figure 2.9 for simulations. The comparisons between the simulation and measurement are given in Figure 2.10. The simulations can reproduce the experimental measurements from the test vehicle. The

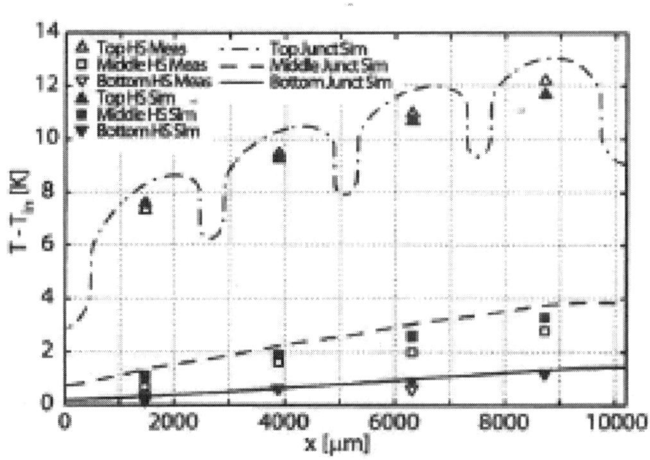

FIGURE 2.10 Simulation and measured data for four hot spots in top layer.

junction temperature needs to be distinguished from the hot spot temperature. The former is the maximum temperature along streamwise directed line, in the symmetry plane, lying between silicon slab and hot spot layers. The latter is the line averaged temperature on the centerline along y axis in Figure 2.9 lying in the central plane of the hot spot and therefore will be lower than the junction temperature.

Cautions must be taken in the design of the multilayer integrated water-cooled chips to prevent any water leakage in the package. The short circuit is developed while in contact with water. From an electric performance point view, the dielectric fluids should be employed in all electric systems. Unfortunately, the thermal performance of the dielectric fluids is much less effective than that of using water, which must be treated in order to be used in the cooling system.

2.2 PRINTED CIRCUIT BOARD

The next level of packaging is the circuit board. The printed circuit board (PCB) which is also often referred to as the printed wiring board (PWB) is a major element of any electronic equipment. This is because the PCB provides a mounting surface and structural support for the electronic components (or chip packages) and wiring channels to serve as conduits for chip-to-chip connections. The PCB is also frequently considered as the primary field replaceable unit and a test bed to provide accessible points in making circuit checks.

2.2.1 Construction

A typical printed circuit board consists of a multilayer of dielectric material and copper film, as shown in Figure 2.11. The copper layers are used to provide the chip-to-chip connections for transmitting signals, power, and grounding purpose. The basic concern in the thermal analysis is how to model the board accurately. Although the majority of material of a PCB (probably over 95% by volume) are insulation materials which have a very low thermal conductivity, one can never omit the contribution due to the copper layers in the thermal analysis. Any omission of the copper layer in the analysis will result in excessive high component

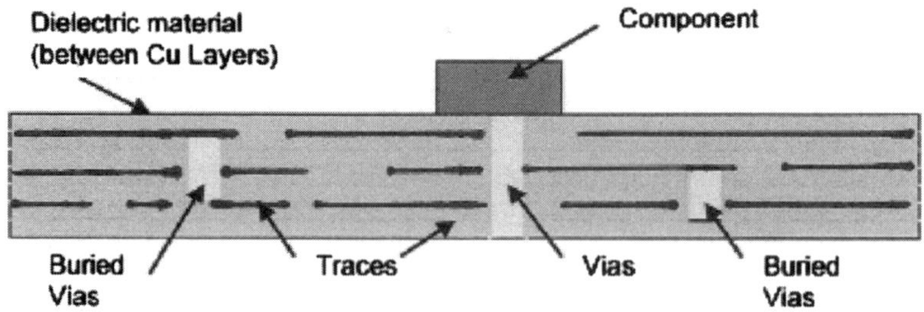

FIGURE 2.11 Typical construction of a PCB.

temperatures, especially for the case of conduction-cooled PCB. The thermal conductivity of the cooper and dielectric materials are listed in Table 2.1.

2.2.2 Equivalent Thermal Conductivity

It is not practical to model the board exactly as shown in Figure 2.11. Therefore, the simplified configuration given in Figure 2.12 is considered. The equivalent thermal conductivities of a printed circuit board (PCB) can be determined in the following ways [3].

Planar Thermal Conductivity, K_{xy}

The laminated layers are considered as parallel paths in calculation of the equivalent thermal conductivity. For a parallel network, the total resistance is

$$\frac{1}{R_{xy}} = \frac{1}{R_1} + \frac{1}{R_2} + \frac{1}{R_3} + \ldots \tag{2.4}$$

where

R_i = thermal resistance for the ith layer, i = 1, 2, 3, ...

$= \dfrac{L_l}{k_i A_i c_i}$

L_i = length of the ith layer
A_i = cross-sectional area of the ith layer (normal to heat flow)
k_i = thermal conductivity of the ith layer
c_i = fraction of the total coverage for the ith layer
(e.g., c = 0.8 if the cut-out for a given layer is 20%)

Rearranging Equation (2.9), we have

$$\frac{K_{xy}A}{L} = \Sigma\left(\frac{c_i k_i A_i}{L_i}\right)_{\text{in}} + \Sigma\left(\frac{c_i k_i A_i}{L_i}\right)_{\text{cu}} \tag{2.5}$$

Table 2.1 Thermal Conductivity of Basic Materials for PCB

Material	W/m °C	But/hr-ft-°F
Copper layer	386.00	223.180
Dielectric layer		
G10	0.30	0.173
FR-4	0.35	0.202
Polyimide	0.52	0.301

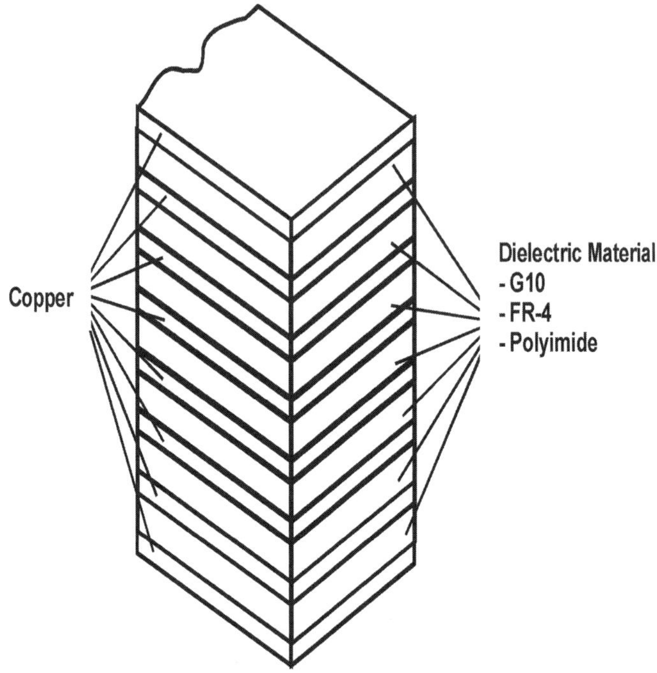

FIGURE 2.12 Simplified board construction for thermal model.

where "in" and "cu" represent the dielectric insulation material and the copper, respectively. A is the total cross-sectional area over the thickness of a PCB, i.e., $A = t$ (thickness) w (width). Since $L = L_1 = L_2 = L_3 = \ldots,$

Equation (2.5) can be rewritten as follows:

$$K_{xy} = \sum\left(\frac{c_i k_i A_i}{A}\right)_{\text{in}} + \sum\left(\frac{c_i k_i A_i}{A}\right)_{\text{cu}}. \tag{2.6}$$

Since c_i is equal to one for the insulation layers, Equation (2.6) can be further simplified to the following form:

$$K_{xy} = k_{\text{in}}\left(\frac{t'_{\text{in}}}{t_{\text{pcb}}}\right) + k_{\text{cu}}\left(\frac{t'_{\text{cu}}}{t_{\text{pcb}}}\right) \tag{2.7}$$

where

t'_{in} = total effective thickness of the dielectric material
$= t_{\text{pcb}} - t'_{\text{cu}}$
t'_{cu} = total effective thickness of the copper film
$= \Sigma\,(c_i\, t_c)_{\text{cu}}$
t_{pcb} = total thickness of the printed circuit board,
$= t_{\text{in}} + t_{\text{cu}}$.

Normal Thermal Conductivity, K_z

Heat flow is normal to each layer, and it is assumed that the resistances are in series as follows:

$$R_z = R_1 + R_2 + R_3 + \ldots \tag{2.8}$$

or

$$\frac{t_{\text{pcb}}}{K_z A_z} = \sum\left(\frac{t_i}{c_i k_i A_i}\right)_{\text{in}} + \sum\left(\frac{t_i}{c_i k_i A_i}\right)_{\text{cu}}. \tag{2.9}$$

Since $A_z = A_i$, the equivalent thermal conductivity in the z-direction is

$$K_z = \frac{t_{\text{pcb}}}{\sum\left(\frac{t_i}{c_i k_i}\right)_{\text{in}} + \sum\left(\frac{t_i}{c_i k_i}\right)_{\text{cu}}}. \tag{2.10}$$

Again, since c_i is equal to one for insulation layers, we have

$$K_z = t_{\text{pcb}} / (t'_{\text{in}} / k_{\text{in}} + t'_{\text{cu}} / k_{\text{cu}}). \tag{2.11}$$

Graebner [7] made detailed measurements of the conductivities for the planar and the normal directions of the board. The conductivities for copper and glass-epoxy are 350 and 0.59 W/m °C, respectively. The board averaged conductivities are correlated into the following equations

$$K_{xy} = 350\,(t'_{\text{cu}} / t_{\text{pcb}}) + 0.8 \tag{2.12}$$

and

$$K_z = [1.69\,(1 - t'_{\text{cu}} / t_{\text{pcb}}) + 0.0026\,(t'_{\text{cu}} / t_{\text{pcb}})]^{-1} \tag{2.13}$$

Later, Graebner and Azar [8] presented empirical equations given below to estimate the thermal conductivities of a PCB. It should be noted that the value of 385 is the thermal conductivity of copper.

$$K_{xy} = 385\,(t'_{\text{cu}} / t_{\text{pcb}}) + 0.87 \tag{2.14}$$

and

$$K_z = [3.23\,(1 - t'_{\text{cu}} / t_{\text{pcb}}) + 0.0026\,(t'_{\text{cu}} / t_{\text{pcb}})]^{-1} \tag{2.15}$$

where the unit of thermal conductivity is W/m °C.

It should be noted that all of the above equations do not include the effects due to the vias and any local non-uniform conditions. The above equations simply consider the copper content in the board.

Table 2.2 Board Construction Information for a 14 Layers Board

Input layer	Input t_{in} (in)	Input t_{cu} (in)	Input cu coverage	Net t'_{cu} (in)
1	0.0036	0.0022	0.28	0.000616
2	0.002	0.0013	0.61	0.000793
3	0.0031	0.0013	0.66	0.000858
4	0.004	0.0006	0.05	0.00003
5	0.0031	0.0006	0.03	0.000018
6	0.004	0.0013	0.62	0.000806
7	0.0031	0.0006	0.05	0.00003
8	0.004	0.0006	0.03	0.000018
9	0.0031	0.0013	0.63	0.000819
10	0.004	0.0006	0.04	0.000024
11	0.0031	0.0006	0.02	0.000012
12	0.002	0.0013	0.66	0.000858
13	0.0036	0.0013	0.61	0.000793
14		0.0022	0.23	0.000506
sum	0.0427	0.0158		0.006181

Notes:
t_{pcb} (in) = 0.0427 + 0.0158 = 0.0585
t'_{in} (in) = 0.058 – 0.006181 = 0.052319

Example 2.1

To compute the equivalent thermal conductivities for K_{xy} and K_z with the input listed in Table 2.2. Table 2.3 includes the comparisons with the present predictions which are based on Equations (2.6) and (2.10) and the results obtained from Equations (2.12) through (2.15). The results from Equations (2.6) and (2.10) compare extremely well with those obtained from the above empirical equations (Equations (2.13) to (2.15)).

Recently, several PCB thermal design tools are available. One of them is ANSYS TASPCB. The users can directly import all the relevant design data from the ECAD tool, including component data, traces, power and ground planes and vias. A built-in component library provides 3D geometry and thermal data to automatically create a 3D board model which is then solved thermally with aid of CFD analysis. The software can actually predict the local thermal conductivities at all three directions at any point of the board.

Table 2.3 Comparisons of Board Thermal Conductivities

	(K_{cu} = 385 K_{in} = 0.3)		(K_{cu} = 350 K_{in} = 0.59)	
	Present	Graebner/Azar (8)	Present	Graebner (7)
K_{xy} (W/m–°C)	40.964668	41.54838	37.508	37.78034
K_z (W/m–°C)	0.335411	0.346141	0.65957	0.6615

2.2.3 Thermal Consideration

One of the critical functions of the board is to spread the heat generated from the components. In fact, the board is the largest heat sink for all components on the board in the case of air cooling. To enhance heat conduction across the board thickness, thermal vias are often utilized. There are two types of the thermal vias. One is unfilled and another is filled thermal via. However, the thermal advantage of the filled thermal via over unfilled one is limited.

The thermally enhanced PCB has been available for a long time. However, this type of the PCB is generally applied to the power converters, motor controls and LED lighting and is available from Laird Technologies and Bergquist.

REFERENCES

1. Bezuk, S., "Flip Chip Challenges", Kyocera America Inc.
2. Yuan, T.D., Pan, H.-Y., Li, Y., "Thermal Interface Material (TIM) Design Guidance for Flip Chip BGA Package Thermal Performance", Altera, San Jose, CA, 2004.
3. Yeh, L.T., and Chu, R.C., Thermal Management of Microelectronic Equipment, ASME Press, 2002.
4. Ramakrishna, K., and Lee, T.-Y.T., "Evaluation of Thermal Enhancements to Flip-Chip-Plastic Ball Grid Array (FC-PBGA)", Journal of Electronic Packaging, Vol. 129, 2004.
5. Alfieri, F., Tiwari, M.K., Zinovik, I., Poulikakos, D., Brunschwiler, T., and Michel, B., "3D Integrated Water Cooling of A Composite Multilayer Stack of Chips", Proceedings of 14th International Heat Transfer Conference, Washington, DC., USA, August 8–13, 2010.
6. Tuckerman, D.B. and Pease, F.F., "High Performance Heat Sinking for VLSI", IEEE Electron Device Letter EDL-2, 1981.
7. Graebner, J.E., "Thermal Conductivity of Printed Wiring Boards", Electronic Cooling, Vol. 1, No. 2, 1995.
8. Graebner, J.E., and Azar, K., "Thermal Conductivity Measurements in Printed Wiring Boards", Journal of Heat Transfer, Vol. 119, 1997.
9. Lee, T.-Y. Tom and Mahalingam, M., "Thermal Limits of Flip Chip Package-Experimentally Validated, CFD Supported Case Studies", IEEE Transaction on Components, Packaging, and Manufacturing Technology, vol. 20, No. 1, 1997.

Chapter 3

FANS AND AIR HANDLING SYSTEMS

The fan is the heart of any forced air cooling system. The fan propeller converts torque from the motor to increase the static pressure across the rotor of the fan and to energize the air particles. A fan is frequently employed in electronic equipment to provide the airflow required for cooling electronics. Direct air cooling is a primary example of using a fan for thermal control of electronic systems.

3.1 FAN TYPES

Fans are generally classified into two types, namely, centrifugal (blower) and axial fans, depending on the direct of airflow through the impeller.

1. Centrifugal Fans
   - receive air at its axis of rotation and exhaust air at its periphery in a direction normal to the rotation axis;
   - produce pressure from two sources (i) centrifugal force created by rotating air column and (ii) kinetic energy imparted to air as it leaves the impeller; and
   - create low flow rate against a high resistance (pressure).
2. Axial Fans
   - deliver air in a direction parallel to the fan blade axis;
   - produce pressure from the change in velocity passing through the impeller; and
   - provide a high airflow rate but tend to work against low pressure.

The axial fan can further be divided into three categories as follows:

a. Propeller fans are the simplest type and are designed to move large volumes of air at low velocity and develop low static pressure.
b. Tube axial fans are similar to the propeller fan but provide a higher total pressure capability than the propeller fans.
c. Vane axial fans are the same as the tube axial but have vanes to straighten the swirling flow created as the air is accelerated which results in improved efficiency and noise characteristics.

Basically, tube axial and vane axial fans fill the requirements between centrifugal and propeller fans.

3.2 FAN PERFORMANCE AND FAN LAWS

The fan performance or characteristic curve is generated through a series of tests as illustrated in Figure 3.1 [1]. The fan is tested from the shutoff conditions to nearly free delivery conditions. At the shutoff point, the duct connected to the fan outlet is completely blocked. On the other hand, the outlet resistance of the duct is reduced to zero at the free delivery point. Additional test points between these two limiting conditions are obtained when various flow restrictions are placed at the end of the duct to simulate various operating conditions on the fan.

Fans designed to be used with a duct are generally tested with a length of duct between the fan and the measurement station. The purpose of this length of duct is to smooth out the flow as it exits the fan and to provide stable and uniform flow conditions at the measurement station. The measured pressures are then corrected back to the fan outlet conditions. Fans designed to be used without a duct are tested without the duct.

As air moves through a system, the pressure is decreased because of expansions, contractions, bends, other structure blockages, and friction. For a system under consideration, the pressure drop is a function of the flow rate, therefore, a system resistance curve can be generated. The intersection of the fan performance and the system resistance curves is the operation point of the system, as given in Figure 3.2 [2].

Figure 3.2 also shows the static efficiency, ($\eta = Qp/P$), which is defined as the ratio of the static pressure power to the input power of a fan. The maximum static

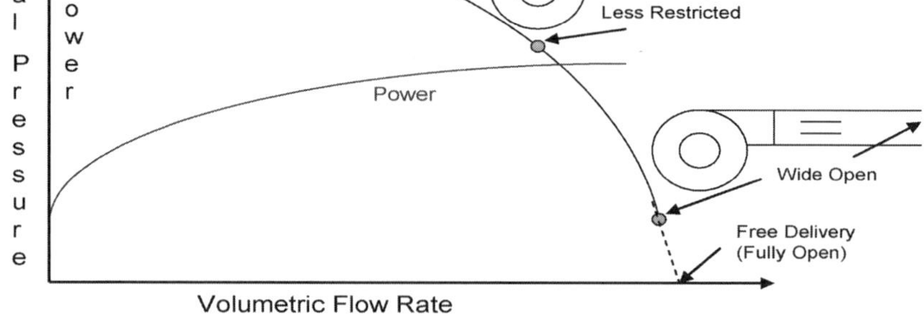

FIGURE 3.1 Method of obtaining fan performance curves.

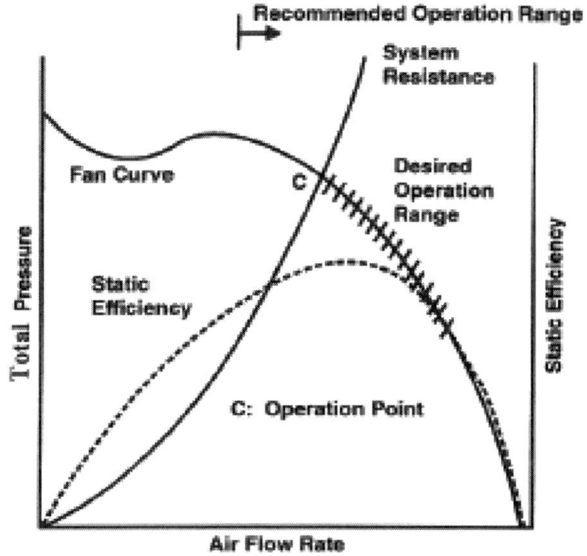

FIGURE 3.2 Typical fan curve and system operation point.

efficiency identifies the ideal combination of the flow rate and pressure of the fan. Fans should be operated as close to this point as possible if they are to achieve optimum results as given in the figure.

The fan laws relate the performance variables for any dynamically similar series of fans. The variables involved are fan size (D), rotational speed (N), gas density (ρ), volumetric flow rate (Q), pressure (p), power (P), and efficiency (η). The relationships among the variables are as follows:

$$Q = \varphi N D^3 \quad \text{or} \quad Q1 = Q2(D1/D2)^3\,(N1/N2) \tag{3.1}$$

$$p = \psi \rho N^2 D^2 \quad \text{or} \quad p1 = p2(D1/D2)^2\,(N1/N2)^2 \tag{3.2}$$

$$P = \lambda \rho N^3 D^5 \quad \text{or} \quad P1 = P2(D1/D2)^5\,(N1/N2)^3 \tag{3.3}$$

where φ, ψ, and λ are the constants for geometrically and dynamically similar operation and are also referred to as the flow coefficient, pressure coefficient, and power coefficient, respectively.

3.3 FAN INLET AND EXIT CONDITIONS

Fan performance is strongly depending on the flow conditions at the inlet and exit of the fan. The air must enter the fan housing as uniformly as possible in order to achieve a better fan performance. In addition, the entering air also requires a sufficient length to reach the developed flow. Therefore, the fan blade within the housing of an axial fan is preferred being located at near the exit end of the fan. Example of the effect of the blade location on fan performance is illustrated in Figure 3.3 [3]. Figure 3.4 shows the effect of the blade location on the fan performance when the fan is located adjacent to a restricted area such as a 90° bend [3].

Hill and Hill [5] have studied experimentally the effects of electronic enclosure on fan performance. Two axial fans with the same speed (3100 RPM) were tested. The fan characteristics are presented in Table 3.1.

The ratio of plenum depth, P to fan tip shroud diameter, D was selected as the parameter for comparison. In addition, the flow rate was normalized based on the free flow rates listed in Table 3.1. The plenum wall has various opening areas. The test results for the 35% plenum area opening are given in Figure 3.5. The findings from the tests can be summarized as follows:

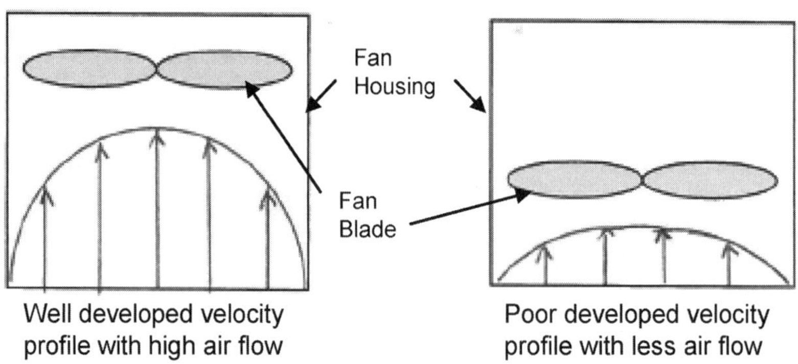

FIGURE 3.3 Effect of blade location on fan performance.

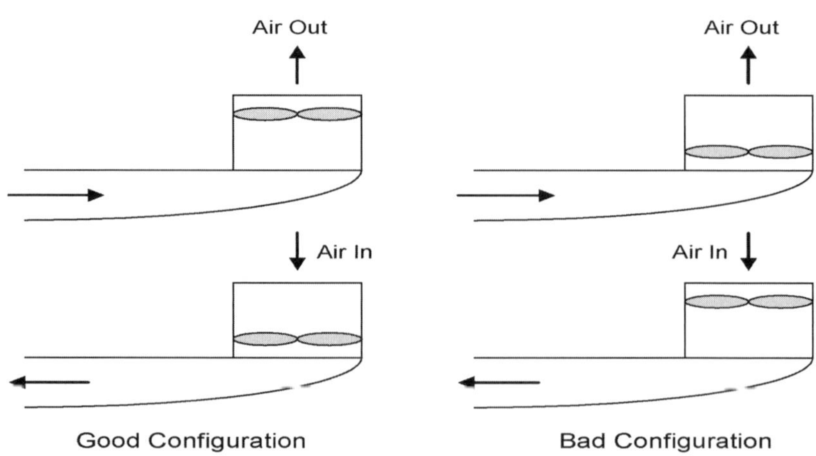

FIGURE 3.4 Effect of blade location on fan performance with fan adjacent to restricted area.

Table 3.1 Characteristics of Tested Fans

	Fan A	Fan B
Tip shroud diameter, *D* (mm)	77.1	112.9
Hub diameter, *d* (mm)	36.1	52.3
Thickness, *t* (mm)	25.0	38.9
Speed, *N* (RPM)	3100	3100
Input power (W)	2.2	5.6
Max. static pressure, Ps (mm wg.)	4.09	7.11
Free flow, (L/s)	17.7	47.5

1. The evacuating flow direction is more susceptible to degraded fan performance at low *P*/*D* ratios than the pressurizing flow direction.
2. Larger fan (Fan B) is not affected by the lower *P*/*D* ratios as much as smaller fan (Fan A) for either flow direction.
3. The results reveal that *P*/*D* ratios greater than 1 do not yield significantly better fan performance than a *P*/*D* ratio of unity.

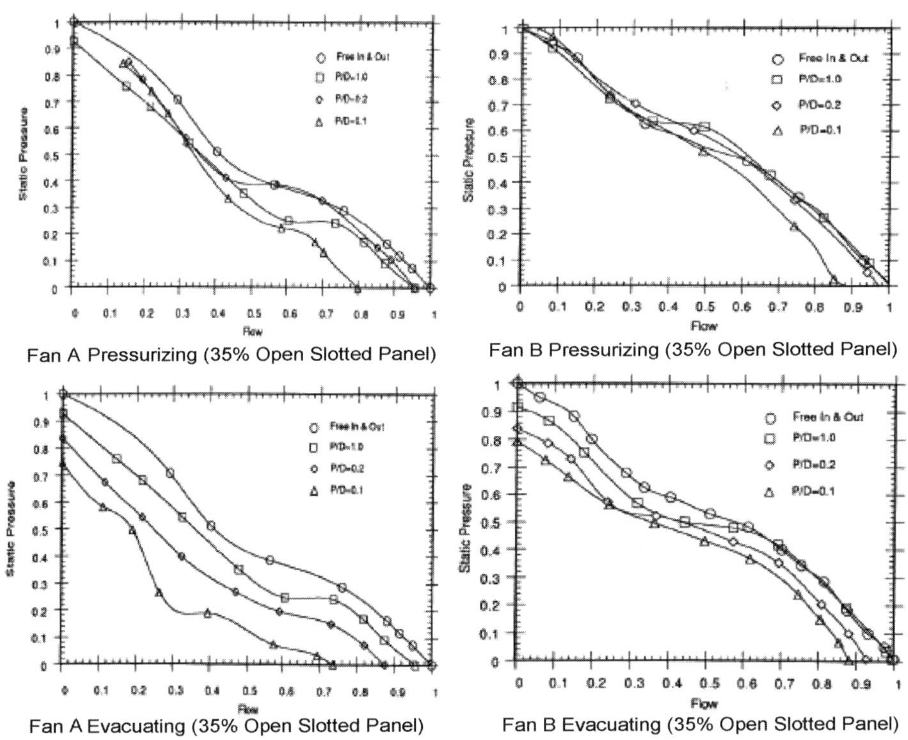

FIGURE 3.5 Fan performance curves for various plenum depth ratio (P/D).

3.4 FAN AND SYSTEM PRESSURE

Fan total pressure is the true indication of the energy to the airstream by a fan. Therefore, the total pressure should be used for the fan selection and air distribution system design. The system pressure loss is defined as the summation of all the individual total pressures in the entire air distribution system.

The system resistance is the sum of various pressure drops in the air delivery system. In addition to the friction loss, the system pressure drop is caused by other losses such as inlet, exit, change of cross section (expansion and contraction), turns, and elbows. The pressure loss in the above items is typically expressed by the velocity head in the following form.

$$\Delta p = k\rho V^2/(2g_c) \tag{3.4}$$

where k is the loss coefficient, and its value for various configurations can often be found in the handbook of hydraulics [4]. g_c is 32.2 lb ft/lb$_f$ sec^2.

Frequently, the pressure drop in an airflow system is expressed in the unit of "in of water." Equation (3.4) for dry air at standard conditions (14.696 psia and 70 °F with air density of 0.075 lbm/ft^3) can be rewritten for this unit as follows

$$\Delta p = k\,(V/4005)^2 \tag{3.5}$$

where Δp is "in of water," and V is ft/min.

If air is not at the standard conditions, Equation (3.5) becomes as follows:

$$\Delta p = k\rho(V/1097)^2 \tag{3.6}$$

The density of air is a function of the pressure and temperature which in turn is a function of the altitudes.

3.5 FAN ACOUSTIC NOISE

Acoustic noise which is more or less a purely subjective personal phenomena has two basic characteristics. One is the physical phenomenon which can be measured, and another one is the psychoacoustic characteristics which are to judge the effect of noise on human beings. The acoustic noise can be measured in terms of "sound pressure" and "sound power" which are two distinct but commonly confused characteristics of sound. However, both share the same unit of measure, the decibel (dB).

"Sound pressure," which is measurable, is a pressure disturbance in the atmosphere whose intensity is influenced not only by the strength of the source but also by the surroundings and the distance from the source to the receiver. The sound pressure is what the ears hear and what the sound meters measure. On the other hand, "Sound power" is the acoustic energy emitted by the sound source. The sound power is an absolute value and is not affected by the environment. The sound power cannot be measured directly and is calculated from the sound pressure measurement.

Fan noise is a function of the fan design, volume flow rate, total pressure, and efficiency. The most efficient operating range for a given line of fans is generally the quietest. Fan noise is expressed by the sound power levels in eight octave bands. It should be noted that low exit velocity does not necessarily guarantee quiet operation. Furthermore, noise comparisons of different types of fan or fans from different manufacturers cannot be made by comparing the fan rotational or tip speeds, and the only valid basis for comparison are the actual sound power levels generated by fans.

The fan noise will be affected by the system in which fan operates. Maling and Yeager [6] provide the following guideline to minimize the fan noise

1. No obstruction should be located within one fan diameter from the fan inlet for axial fans so that the airflow entering the fan inlet is as spatially uniform as possible.
2. Pulling air over equipment being cooled usually causes undesirable turbulence at the (exhaust) fan inlet and increases noise level.
3. Designing the system to have the lowest possible static pressure rise for the required airflow so that the fan can operate at low tip speeds and reduce the noise level.

3.6 FAN RELIABILITY

The fan is the heart of any forced air cooling system and therefore has been a main contributor to the failure of many electronic systems. Just as microelectronics, fan failure can be classified into "catastrophic failures" in which fan is not running due to hardware failures and "software failures" in which fan can still be functioning but fails to meet performance requirements. Since a fan consists of both mechanical and electrical parts, the major fan failure modes as stated by Tian [7] include mechanical failures, electrical failures, failures due to installation, acoustic failures, failures due to poor serviceability, and poor cooling performance failures and the pareto of the fan failures is given in Figure 3.6.

Fans can fail in many ways, and failures can also be defined differently by the different vendors and/or applications. Two metrics of fan reliability commonly quoted by vendors are the L2 life and L10 life which are the second and tenth percentiles under some assumed fan life distribution. In other words, L2 and L10 indicate the time points at which 2% and 10% of the fans will fail, respectively. The L2 and L10 can be determined by the following equations [8]

$$L2 = \alpha\,(0.02020)^{1/\beta} \tag{3.7a}$$

$$L10 = \alpha\,(0.10536)^{1/\beta} \tag{3.7b}$$

where α is the characteristic life, and β is the shape parameter.

The shape parameters for Weibull model are generally greater than 1, which means that a fan failure increases with age. Reference 7 indicates in the high reliability/

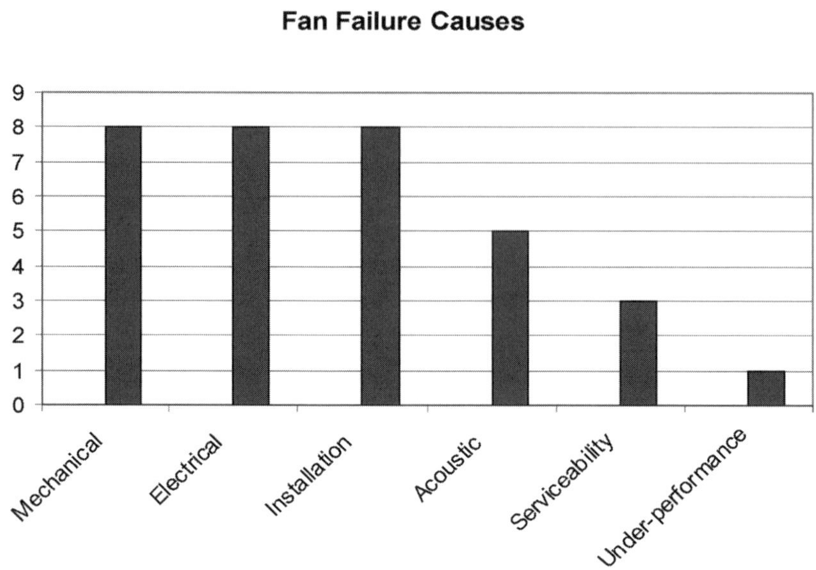

Figure 3.6 Causes of fan failure.

availability products, such as servers, a more useful metrics is $L1$ life that means the life when 1% of products fail. $L1$ can be expressed as follows:

$$L1 = \alpha(0.01005)^{1/\beta} \tag{3.7c}$$

Vendors also often use the mean time to failure (MTTF) as the reliability indicator. For the Weibull distribution, MTTF can be expressed as follows:

$$\text{MTTF} = \alpha\Gamma(1 + 1/\beta) \tag{3.8}$$

where Γ denotes the Gamma function.

It is often confused MTTF with MTBF (mean time between failures). It should be noted that the MTBF should only be used in a repairable system setting in order to understand the system's maintenance needs and the service cost.

The life of most fans is generally limited by the bearing life because the bearing failure is the major failure mechanism in most fans. The bearing life is strongly depending on the composition and reliability of the bearing lubricants (e.g., greases). Kim and Claassen [8] recommend to limiting the bearing temperature rise to 10°C or less where the fan reliability is critical. The authors further recommend the Booser grease life equation as given below for the fan life estimation:

$$\log L10 = 2.6 + (\text{Kt/Tbrg}) - 0.301\,\text{S} \tag{3.9}$$

where

S	=	$Sg + Sn + Sp$
Sn	=	$0.86\ DN/(DN)_L$
Sp	=	$0.61\ DNP/Cr^2$
P	=	Equivalent dynamic bearing load, lbf
N	=	Speed, rpm
Cr	=	Basic dynamic load capacity, lbf
D	=	Bore diameter, mm
$(DN)_L$	=	Speed limit, rpm-mm
S	=	Half-life subtraction factor; for S = 1, the life drops 50%
Sg	=	Grease half-life subtraction factor; typically 0 for many greases
Sn	=	Speed half-life subtraction factor
Sp	=	Load half-life subtraction factor
Kt	=	Grease temperature factor = 2450 for acceleration factor of 1.5 for each 10°C
Tbrg	=	Bearing temperature, K

It should be noted that Equation (3.9) does not take into account of the effect of grease quantity and may not cover all greases on the market, especially for those newly developed greases which use synthetic oils.

REFERENCES

1. Fans, Chapter 3, ASHRAE 1983 Equipment Handbook, 1983.
2. Yeh, L. T., and Chu R. C., Thermal Management of Microelectronic Equipment, ASME Press, New York, 2002.
3. Steinberg, D. S. Cooling Techniques for Electronic Equipment, Wiley, New York, 1980.
4. Belevins, R. D., Applied Fluid Dynamics Handbook, Van Nostrand Reinhold, New York, 1984.
5. Hill, T. B. and Hill, C. C., "Effects of Electronic Enclosure Layout on Fan Performance," 90-WA/EEP-6, ASME Winter Annual Meeting, Dallas, Texas, November 25–30, 1990.
6. Maling G. C., and Yeager, D. M., "Minimizing Acoustical Noise in Electronic Systems," Electronics Cooling, Vol. 2, No. 2, May 1996.
7. Tian, X., "Cooling Fan Reliability: Failure Criteria, Accelerated Life Testing, Modeling and Qualification," Itherm 2006.
8. Kim, S. and Claassen, A., "How to Evaluate Fan Life," Electronics Cooling, Vol. 2, No. 2, 1996.

Chapter 4

AIR COOLING AND HEAT SINKS

Simplicity and easy maintenance make direct air cooling a most attractive approach in cooling of electronics. This approach includes both active and passive methods. The former is related to forced convection while the latter represents free or natural convection. The passive cooling has always been the preferred choice. However, it is generally limited to the systems with very low heat dissipation because of poor heat transfer coefficients involved in the free convection.

Forced air cooling over printed circuit boards (PCBs) is often encountered in electronic equipment. Various component size and spacing cause flow separation over components and recirculation between components. The flow field is a fully three-dimensional (3D) problem and extremely complex. Heat transfer involves convection from components and the PCB to the air stream, conduction from components to the PCB and heat spreading along the PCB, and also radiation exchange among components and from components to surroundings.

4.1 NATURAL CONVECTION

The natural convection over a vertical plate is most simple and is also often encountered in the engineering applications. To enhance heat transfer, the fins are frequently added to the surface that is also often referred to as the heat sink. Figure 4.1 shows a typical configuration of a continuous fin array. The flow field over a finned heat sink is much complicated than the typical flow over a single plate or in parallel plates because of the involvement of the third surface (fin base).

There are many studies of the natural convection from the multiple surfaces such as parallel plates or finned heat sinks by numerical analysis or by experiment. The finned heat sink as presented in Figure 4.1 consists of a number of U-shaped channels. The corner regions formed by the base plate and the fins result in a significant reduction in the heat transfer.

For vertically straight-fin heat sinks, several experimental data [1–3] are available. Among them, Izume and Nakamure [3] developed a mathematical relationship describing heat transfer from the finned heat sink; however, their equation does not hold in the limiting cases of very large or very small ratios of the channel depth to channel width. To overcome this problem, Van De Pol and Tierney [4] developed the following empirical equation applicable to any channel depth to width ratios.

$$\mathrm{Nu_r} = (\mathrm{Ra}^*/\psi)\{1 - \exp[-\psi(0.5/\mathrm{Ra}^*)^{3/4}]\} \tag{4.1}$$

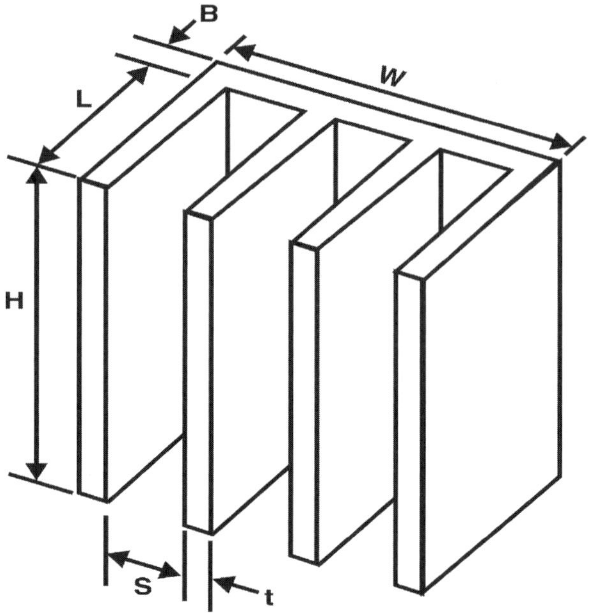

Figure 4.1 A vertical straight fin heat sink.

where

$$
\begin{aligned}
\mathrm{Nu} &= h_u r/k,\\
r &= 2LS/(2L+S),\\
\mathrm{Ra}^* &= (r/H)Gr_r Pr,\\
Gr_r &= g\beta(\rho/\mu)^2\,(T_w - T_a)\,r^3,\\
\psi &= 24(1-0.483e^{-0.17/a})/\{(1+0.5a)[1+(1-e^{-0.83a})(9.14a^{0.5}e^{V*S}-0.61)]^3,\\
a &= S/L,\\
V &= -11.8(1/\mathrm{in}).
\end{aligned}
$$

The above correlation is limited to the constant wall temperature condition and is only applicable to the continuous straight fins. In addition, h_u is the heat transfer coefficient for a U-channel. It should be noted that Equation (4.1) can be applied to various sizes of the heat sink. Radiation heat transfer must always be included under the natural convection conditions. This is especially true for the air at high altitudes where the effectiveness of natural convection is significantly reduced due to reduction of the air density.

Yeh et al. [5] have performed a CFD analysis on the heat transfer from the vertical finned heat sinks in the natural convection environments. The authors also compared the results from the above equation with those of the CFD. A good agreement between both sets of the results further validates the accuracy of Equation (4.1). The equation is considered especially useful for optimizing the heat sink configurations because of its simplicity and ease for obtaining quick results. Yeh et al. [5] further found that the radiation loss is relatively insensitive to fin spacing if the spacing variation is small. Therefore, the natural convection alone can be used to optimize fin spacing of the heat sinks under the natural convection condition.

Yeh [6] utilized the above equation for optimizing component heat sinks. The heat sink widths under consideration are 40 mm, 65 mm, and 90 mm and each width with four different heat sink lengths (30 mm, 40 mm, 50 mm, and 60 mm) in the flow direction. For a given heat sink width and length, analyses are performed to determine the optimal fin spacing for the maximum natural convection heat loss. The fin thickness for all configurations is 1.2 mm, and the heat sink is made of anodized aluminum. The ambient is assumed at 50°C and the heat sink base at 85°C. The former represents inlet air temperature over a printed circuit board, and the latter corresponds to the component case temperature.

It should be noted that all other dimensions of the heat sink remain unchanged while varying the fin spacing to achieve the maximum heat loss which is the product of the heat transfer coefficient, heat transfer surface area, and the overall surface efficiency. The summary results from Yeh [6] are shown in Figures 4.2, 4.3, and 4.4 for the heat sink widths of 40 mm, 65 mm, and 90 mm, respectively. The optimal fin spacing is in theory a function of fin thickness, heat sink width, and length. As shown in Figure 4.2 for the heat sink length of 30mm, its optimal fin spacing is 4.34 mm. On the other hand, the optimal spacing for the heat sink lengths of 40mm, 50mm and 60mm is 5.27 mm. Examining the results for the heat sink length of 30mm in details reveal that the heat loss for the fin spaces of 4.34 mm and 5.27 mm are 4.047 and 4.039 Btu/hr, respectively. The difference in the heat loss between these two fin spaces is so small that one can consider the optimal fin spacing for the 30mm long heat sink to be 5.27 mm in order to reduce manufacturing processes and part counts. Therefore, the optimal fin spacings

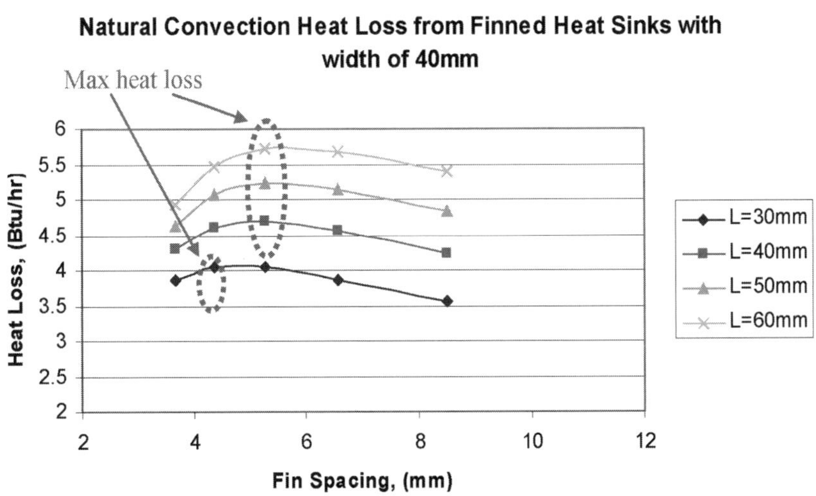

FIGURE 4.2 Natural convection loss for heat sink width of 40 mm.

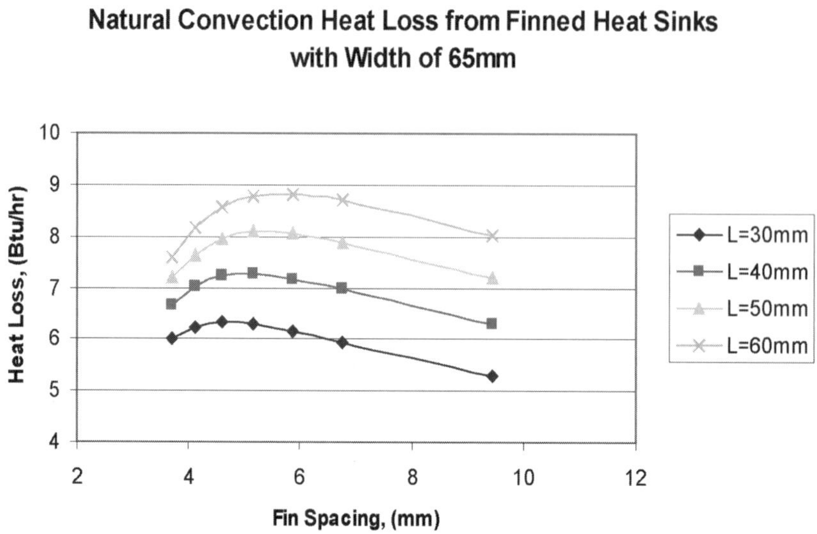

FIGURE 4.3 Natural convection loss for heat sink width of 65 mm.

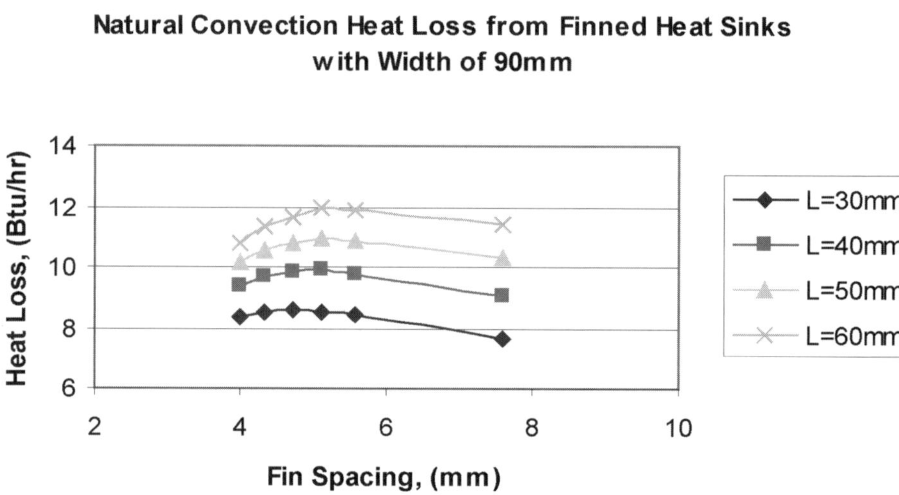

FIGURE 4.4 Natural convection loss for heat sink width of 90 mm.

for heat sink lengths of 30mm, 40mm, 50mm, and 60mm are 5.27mm, 5.18mm, 5.14mm for the heat sink width of 40mm, 60mm and 90mm, respectively. Furthermore, for simplicity and practical application, the single fin spacing of 5.2 mm which is an average value of the above three optimal fin spaces (5.27mm, 5.18 and 5.14mm) is recommended for this family of heat sinks with length ranging from 30mm to 60mm and width from 40mm to 90mm under the natural convection environment.

4.2 FORCED CONVECTION

Simplicity, easy maintenance and low cost make direct air cooling a most attractive approach in cooling of microelectronics. In fact, almost all telecommunication equipment is cooled by air. Various types of heat sinks are often mounted to the components in order to enhance convective heat transfer so that the components can be kept below their respective temperature limits.

4.2.1 Heat Sink Thermal Performance

The presence of the fins on a given surface will affect the local heat transfer coefficient because the flow pattern has been changed. However, one should never just look at the heat transfer coefficient alone in evaluating the thermal performance of a heat sink. It should be noted that the fins generally will increase not only the heat transfer but also the pressure drop. Any thermal design must always have a good balance between the heat transfer and pressure drop. The thermal performance (thermal resistance and pressure drop) of a heat sink can be illustrated as shown in Figure 4.5.

The total thermal resistance of a finned heat sink consists of the average heat sink resistance and the spreading resistance. The former corresponds to the convective heat transfer from the fins, and the latter represents the heat conduction spreading along the base of the heat sink. The total thermal resistance can be expressed as follows:

$$Rt = Ro + Rc \tag{4.2}$$

where

Ro = average heat sink resistance
Rc = heat sink spreading resistance

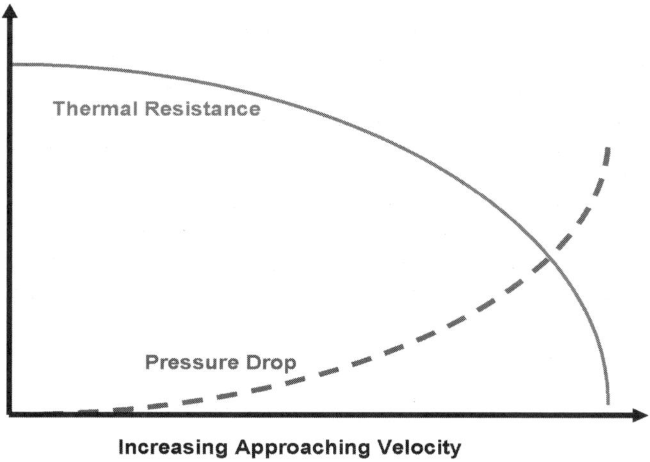

FIGURE 4.5 Typical thermal performance of a heat sink.

Average Heat Sink Thermal Resistance

For a uniform base temperature, the effectiveness of a heat sink is measured by the total heat transfer rate which is defined as follows:

$$Q = \eta_o h A_t \Delta T \tag{4.3}$$

where

Q = total heat transfer rate
η_o = overall surface efficiency
= $1 - A_f/A_t\,(1 - \eta_f)$
A_f = fin surface area
A_t = total heat sink surface area for convection
η_f = fin efficiency
h = heat transfer coefficient
ΔT = temperature difference between average heat sink temperature and fluid temperature

In addition, the average thermal resistance of the heat sink is also given by the following equation:

$$\mathrm{Ro} = Q/\Delta T = 1/(\eta_o h A_t) \tag{4.4}$$

The average thermal resistance of a heat sink is basically related to the fin thermal performance, which is affected by the fin surface area, fin configuration, fin material, and the mass flow rate of the fluid.

Base Spreading Thermal Resistance

The non-uniform temperature distribution along the heat sink base will significantly reduce the thermal performance of a heat sink. Because of the increase in the chip power in the recent years, the oversized heat sinks (larger than the component foot print) are often needed in order to maintain the proper component temperatures. Therefore, the spreading resistance of the heat sink base become a critical issue for cooling of modern microelectronics.

In theory, the thermal spreading resistance can be eliminated if the base thickness or thermal conductivity approaches infinite. The spreading resistance always exists in the practical applications. Song et al. [7] and Lee et al. [8] provided a closed form equation as given below to calculate the thermal spreading resistance.

$$\mathrm{Rc} = [(\mathrm{Ap}^{0.5} - \mathrm{As}^{0.5})/(k\pi^{0.5}\mathrm{Ap}^{0.5}\mathrm{As}^{0.5})] \tag{4.5}$$
$$[\lambda k \mathrm{ApRo} + \tan h(\lambda \mathrm{t})]/[1 + \lambda k \mathrm{ApRo} + \tan h(\lambda \mathrm{t})]$$

and

$$\lambda = \pi^{1.5}/\mathrm{Ap}^{0.5} + 1/\mathrm{As}^{0.5} \tag{4.6}$$

where

As = footprint of heat source
Ap = footprint of heat sink base
t = thickness of heat sink base
k = thermal conductivity of heat sink base
Ro = average heat sink thermal resistance

The graphical presentation of Rc and Ro are illustrated in Figure 4.6. As indicated in the figure, the maximum thermal spreading or constriction resistance Rc, accounts for the local temperature rise over the average surface temperature of the heat sink. It should be noted that the above correlation does not address the shape of the heat source nor the heat sink base.

Later, Lee [9] extended the earlier work to examine the effects of the location and the size of the heat sources. The effect of the heat source size on the heat sink base temperature is shown graphically in Figure 4.7. As shown in the figure, the maximum temperature at the center of the heat sink increases as the size of the heat source decreases. As the temperature rises at the center, the temperature along the edges of the heat sink decreases simultaneously. It should be noted that the area-averaged surface temperature of the heat sink remains unchanged. In other words, the area-averaged surface temperature of the heat sink depends solely on the total heat load.

The effect of the location of heat source on the thermal resistance of the heat sink is illustrated in Figure 4.8. The higher thermal spreading resistance in Case B is due to the fact that heat spreading is limited to two directions while heat spreading in Case A is in three direction as marked by the "arrows." The thermal spreading resistance is significantly increased for the oversized heat sinks.

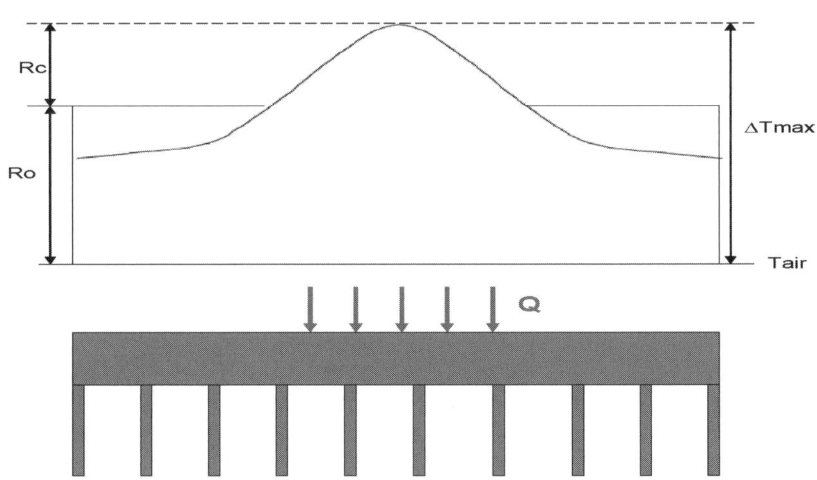

FIGURE 4.6 Schematic view of local resistance or temperature variation of heat sink.

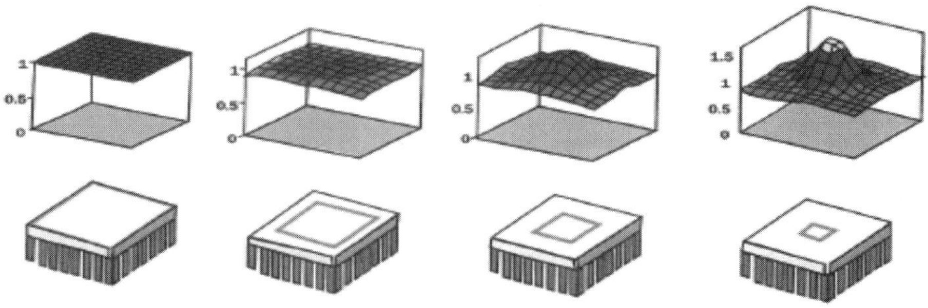

FIGURE 4.7 Normalized local temperature rise with various size of heat sources.

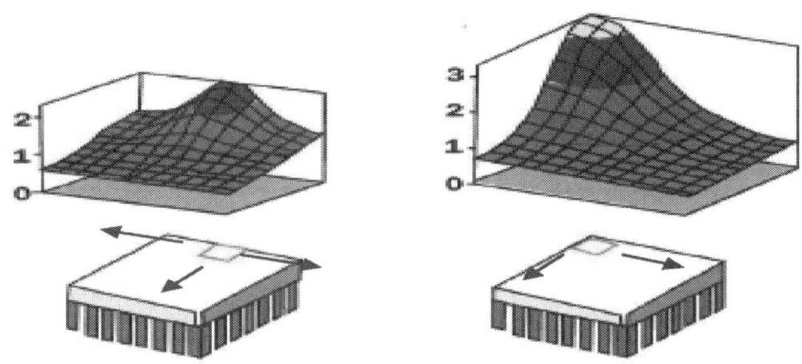

FIGURE 4.8 Effect of heat source location on heat sink thermal resistance.

4.2.2 Factors Affecting Heat Sink Thermal Performance

The following factors will have great impacts on the performance of the heat sink.

Flow Bypass and Leakage

Figure 4.9 shows a typical straight fin heat sink attached to a component on the board. Two important phenomena associated with the air-cooled heat sinks are the flow bypass and the heat sink flow leakage as illustrated in Figures 4.10 and 4.11, respectively. The flow bypass rate equals to the difference between the channel and heat sink inlet flow rates, while the leakage flow rate equals the difference between the heat sink inlet and exit flow rates. Both phenomena will reduce the thermal performance of the heat sink.

No air flow bypass occurs if the cross-section of the test channel is equal to that of the heat sink. However, the flow bypass always exists with air flowing over

Figure 4.9 Typical straight fin heat sink.

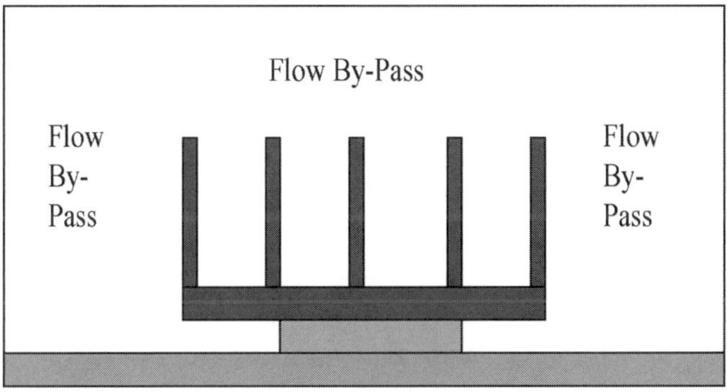

Figure 4.10 Flow bypass over heat sinks.

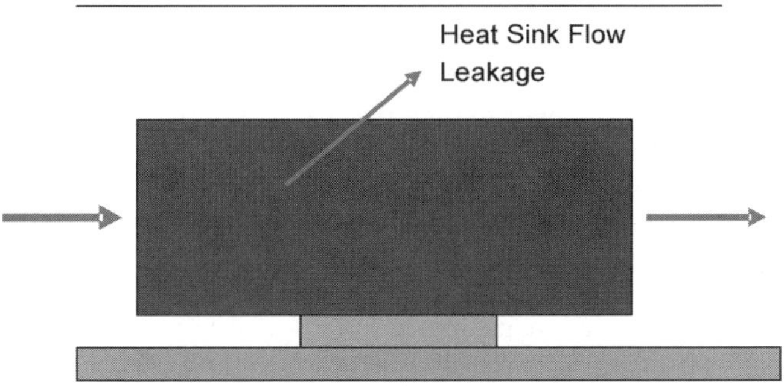

Figure 4.11 Heat sink flow leakage.

a printed circuit board. The main reason for the flow bypass is that the air will flow through the least resistance path. The heat sink flow leakage also exits. One way to eliminate the heat sink flow leakage is to add a cover over the heat sink which flow is totally within a confined space.

Fin Configurations

The fins should have good aerodynamic characteristic to reduce the flow resistance. Yeh [10a–10c] performed CFD analysis to evaluate the thermal performance

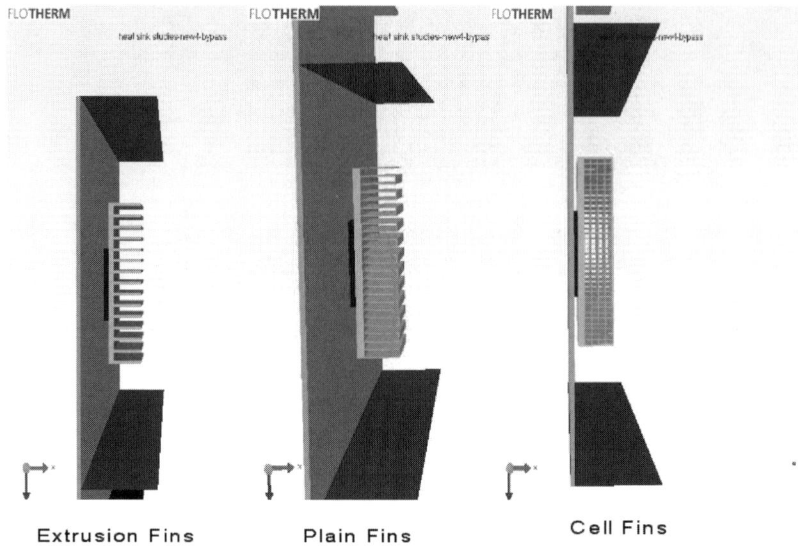

FIGURE 4.12 Various types of heat sinks.

of various types of heat sinks. Three different types of the heat sinks, including extrusion fin, plain fin, and cell fins as shown in Figure 4.12 are under consideration. The dimensions of the fins are given in Table 4.1.

The detailed information of a single fin for the cell fin heat sink is shown in Figure 4.13. All cell fins are enclosed by the frame with the thickness of 0.5 mm.

The case without the flow bypass is first considered. In other words, the individual heat sink is enclosed by the channel walls. The analysis is further extended to the case with the flow bypass. The channel configurations for the heat sinks with the flow bypass are illustrated in Figure 4.14.

The heat transfer coefficient and the thermal resistance of the heat sinks for the case with the flow bypass are presented in Figures 4.15 and 4.16, respectively. The reason for the lowest heat transfer coefficient for the cell fin hat sink is due to the small spacing between fins which results in low air flow through the heat sink.

Table 4.1 Dimensions of Three Heat Sink Configurations

	Extrusion fin (mm)	Plain fin (mm)	Cell fin (mm)
Length (*L*)	60.0	60.0	60.0
Width (*W*)	75.95	75.95	75.95
Height (*H*)	17.17	17.17	17.17
Base height (H_b)	2.95	2.95	2.95
Fin height (H_f)	14.22	14.22	14.22
Fin thickness (*t*)	1.3	0.2	0.1
Fin number	15	25	35

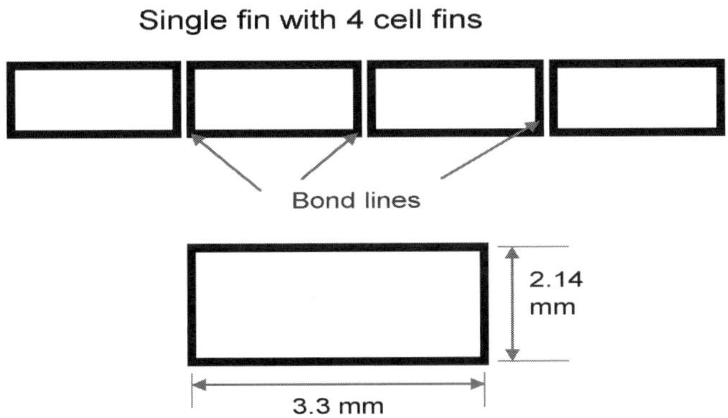

FIGURE 4.13 Configuration of single fin of cell fin heat sink.

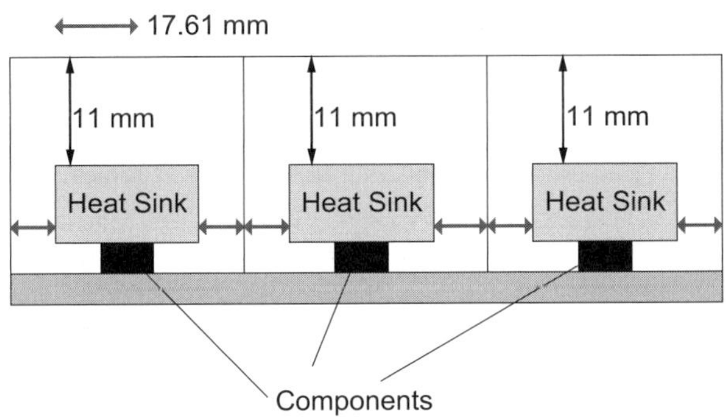

FIGURE 4.14 Channel cross section with flow bypass.

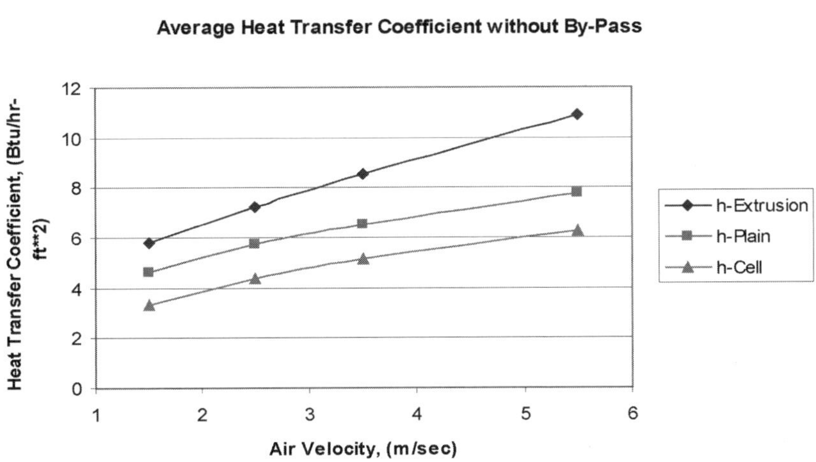

FIGURE 4.15 Average heat transfer coefficient of heat sink without flow bypass.

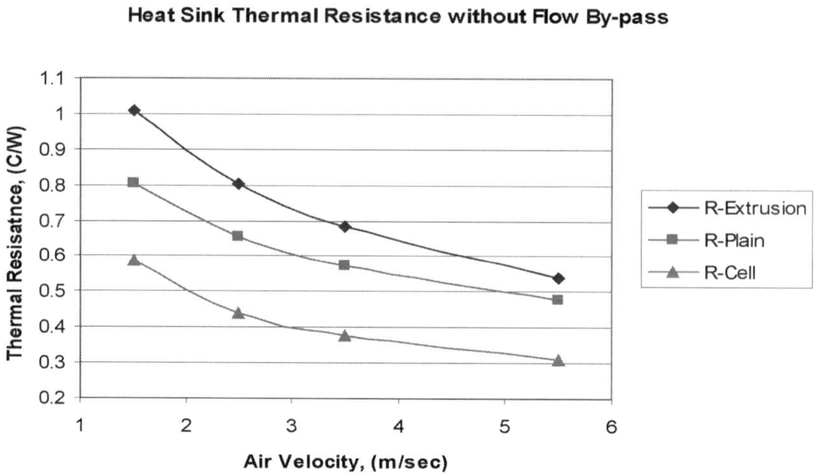

FIGURE 4.16 Average thermal resistance of heat sink without flow bypass.

However, it has the largest heat transfer surface area among all heat sinks under consideration. Therefore, the cell-fin heat sink has the lowest thermal resistance as shown in Figure 4.16.

The heat sink heat transfer surface used to compute the heat transfer coefficient is defined as the total surface area of the heat sink except the bottom of the heat sink base overhanging over the components. The surface areas for the extrusion-fin, the plain-fin, and the cell-fin are 46.74, 73.19, and 140.79 in^2, respectively. The cell-fin heat sink has the largest surface area for the convection, but the space between the fins is very small that it limits its heat transfer coefficient.

The heat transfer coefficient and the heat sink thermal resistance with the flow bypass are presented in Figures 4.17 and 4.18. Generally, the heat transfer coefficients with the flow bypass are less than those without the flow bypass at the same velocities because the less flow goes through the heat sinks.

The trends for the heat transfer coefficient and the thermal resistance are similar to those without the flow bypass with the exception of the low air velocity. As can be seen from Figure 4.18, the thermal resistance of the cell-fin heat sink is across over the plain fin and extrusion fin heat sinks for the velocities less than 2.5 and 1.5 m/sec, respectively. Because the heat transfer coefficient is so small for the cell-fin heat sink at these velocities that the product of the heat transfer coefficient and the surface area is also smaller than those of the other two heat sinks at the same velocity.

The pin fin heat sink has been employed extensively in the direct air cooling for microelectronics. The pin fin heat sink is better suite for the case where the air flow is not parallel to the heat sink. In other words, there is no flow direction limitation for the pin fin heat sink. The local heat transfer coefficient for a pin fin heat sink is generally higher than that of the continuous fin heat sink. However, the latter has a much larger surface area. Therefore, the overall thermal performance

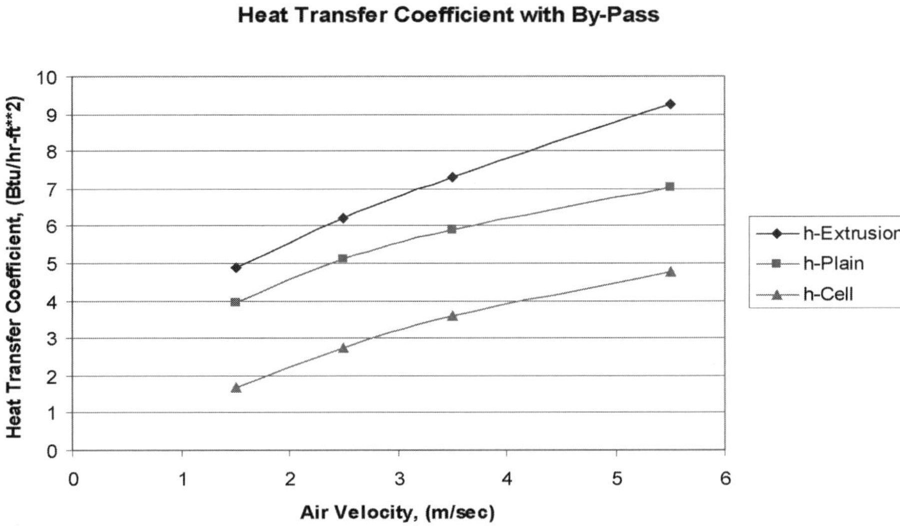

Figure 4.17 Average heat transfer coefficient of heat sink with flow bypass.

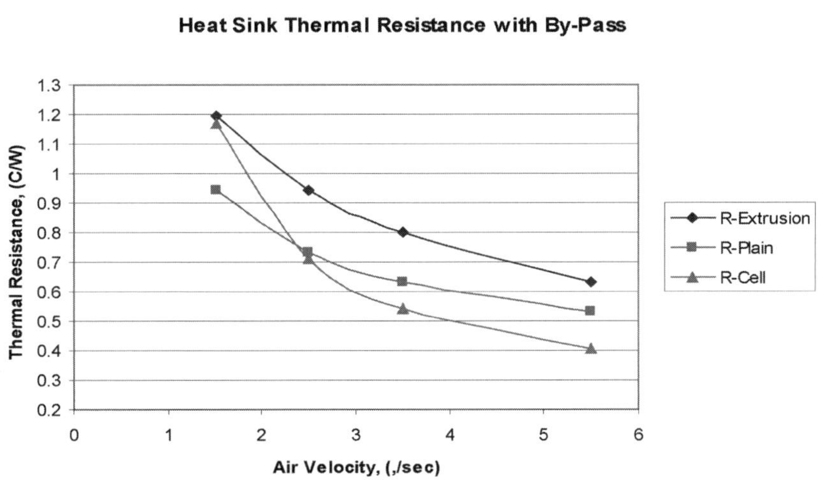

Figure 4.18 Average thermal resistance of heat sink with flow bypass.

of the continuous fin heat sink in general is better than that of a pin fin heat sink. The comparison of thermal resistance between the pin heat sink made by die-casting and the discrete continuous heat sink made by forging process is presented in Figure 4.19 [11]. It should be noted that the thermal conductivity of the forged heat sink is higher than that of the die-casted heat sink.

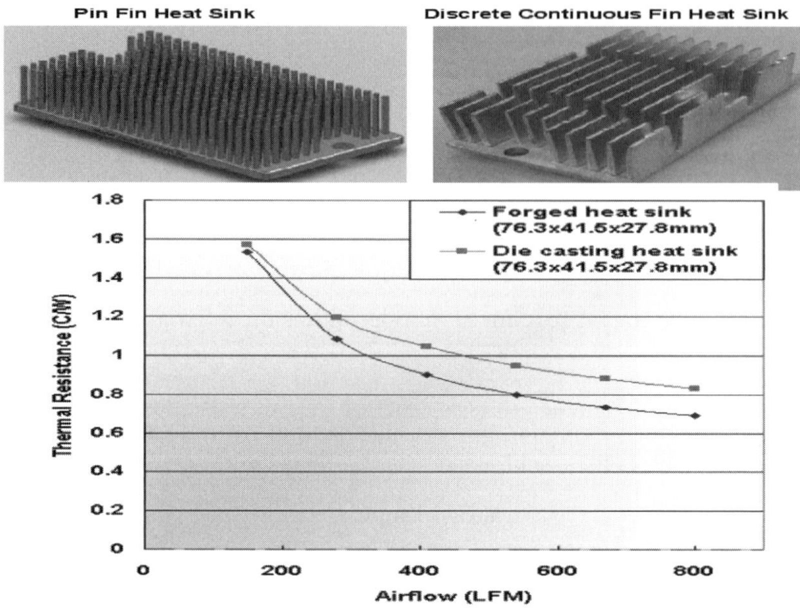

FIGURE 4.19 Thermal resistance for pin fin and discrete continuous heat sinks.

Mass Flow Rates

The heat transfer coefficient is a function of the mass flow rate ($m = \rho VA$) or mass velocity ($G = \rho V$). Generally, the heat transfer coefficient varies with $G^{0.5}$ and $G^{0.8}$ for the fully developed laminar and turbulent flow, respectively. Therefore, the heat transfer coefficient will increase when the mass flow rate is increased. However, on the other hand, the thermal resistance as defined in Equation (4.4) decreases as the mass flow rate increases. In addition, the pressure drop which is proportional to (ρV^2) increases as the mass flow rate increases.

It should be noted that the mass flow rate should be utilized in calculating the heat transfer coefficient or the pressure drop. It is a common mistake by adapting the volumetric flow rate in evaluating the heat transfer or pressure drop. This is especially true for the air cooling at high altitudes where the density is less than that at the sea level because the air density is a function of altitudes. Omission of density in thermal analysis may lead to the equipment overheating at higher altitudes.

4.2.3 Heat Sink Thermal Enhancement

Once the factors affecting the thermal performance of a heat sink are fully understood, thermal enhancements can be made accordingly. The thermal improvements should focus at increasing the total heat transfer rate, not the heat transfer coefficient alone.

Improved Fin Thermal Performance, Ro

As mentioned previously, the average heat sink resistance, Ro is generally related to the fin thermal performance. Various optimizations, such as optimizations for cost or weight/volume or heat transfer, can be made for a heat sink. However, the present discussion is limited to the thermal optimization. The most common one is to optimize the fin spacing or number of fins for the maximum heat transfer rate. A minimum thermal resistance, Ro as defined in Equation (4.4) always existed for a given heat sink as number of fins increasing over a wide range.

The heat transfer coefficient is a function of the mass flow rate. The mass velocity over the individual fins decreases as the number of fins increases. Therefore, heat transfer coefficient alone always decreases with increasing the number of fins due to reduction of air flow over individual fins.

Enhanced Base Thermal Spreading, Rc

High thermal conductivity materials can be used to lower the thermal spreading resistance at the heat sink base. Aluminum heat sink has been widely adapted in the industry because of its light weight and low cost. However, its thermal conductivity is much smaller than copper.

The following methods can be considered to reduce the base thermal spreading resistance which in turns also lowers the total thermal resistance of the heat sink.

1. Copper Embedded Heat Sinks
   The oversize heat sinks are often required due to the increased component power. Aluminum heat sinks are increasingly difficult to meet the thermal requirements of high power components. The hybrid heat sink consisting of aluminum fins and copper base can be employed to reduce the thermal spreading resistance and also to lower the total thermal resistance of a heat sink.

   It is very difficult to measure or even compute the base thermal spreading resistance alone. Therefore, the comparison is often made at the total thermal resistance of the heat sink. McLean [11] has performed several studies and the comparison between the aluminum and copper embedded heat sinks are given in Table 4.2. The improvement is found to be increased with the velocity.
2. Embedded Heat Pipe Base
   The copper heat sink is sometimes not sufficient or too heavy to meet thermal design requirements. Other effective methods such as the embedded heat pipe or vapor chamber base heat sinks. The equivalent thermal conductivity of a heat pipe is typically several times higher than copper.

Table 4.2 Thermal Performance of Aluminum and Copper Embedded Heat Sinks

	AL heat sink 	Copper embedded heat sink 	
Dimension	112 x 74 x 30 mm	112 x 74 x 30 mm	
Heat sink base	7 mm	7 mm	
Heat source	37 x 37 mm	37 x 37 mm	
Power dissipation	80 W	80 W	
	Rt (°C/W)	**Rt (°C/W)**	
LFM	**Thermal resistance**	**Thermal resistance**	**Improvement**
200	**0.537**	**0.522**	**2.7%**
360	**0.366**	**0.351**	**4.1%**
520	**0.316**	**0.295**	**6.5%**
680	**0.283**	**0.258**	**8.8%**
840	**0.264**	**0.236**	**10.6%**

The base of a heat sink can be made of aluminum with the embedded heat pipes. The comparisons between aluminum, embedded copper, and embedded heat pipe heat sinks are presented in Table 4.3 [11].

3. Vapor Chamber Base
   The vapor chamber is similar to the heat pipe which is a two-phase heat transfer device. The vapor chamber offers two district advantages over the heat pipes, and these are the direct contact to the heat source and uniform spreading in all directions. The typical vapor chamber consists of the vapor chamber, a (aluminum) fin stock, and an aluminum frame for mechanical attachment. These three parts are soldered together as shown in Item A in Figure 4.20 [12]. Another option is to directly embed the vapor chambers into the base of the heat sink as given in Item B of Figure 4.20 [12]. Meyer [12] concludes that vapor chambers can offer thermal performance improvement on the order of 10% to 30% over copper base and slightly better than heat pipe base heat sinks. It should be noted that the vapor chamber does include the wick structures as used in the heat pipes.

Recent development efforts are focusing at the new generation of the vapor chambers with the base thickness of the heat sinks no more than 3 mm. The thermal performance of such thin base vapor chamber heat sinks and the embedded heat pipe heat sinks were studied by Glover et al. [13]. The test samples (heat sinks with various types of constructions of vapor chambers) are given in Table 4.4. The overall dimensions of the various heat sinks are 152.4 mm (width) × 101.6 mm (length) (in flow direction) × 12.7 mm (total height). The tests were also conducted at various orientations, including (a) horizontal with bottom heating (fins

Table 4.3 Comparisons of Aluminum, Copper Embedded, and Heat Pipe Embedded Heat Sinks

	AL heat sink 	Copper embedded heat sink 	Heat pipe embedded 
Dimension	112 x 74 x 30 mm	112 x 74 x 30 mm	112 x 74 x 30 mm
Heat sink base	7 mm	7 mm	7 mm
Heat source	37 x 37 mm	37 x 37 mm	37 x 37 mm
Power dissipation	80 W	80 W	80 W
	Rt (°C/W)	**Rt (°C/W)**	**Rt (°C/W)**
LFM	**Thermal resistance**	**Thermal resistance**	**Thermal resistance**
200	**0.537**	**0.522**	**0.516**
360	**0.366**	**0.351**	**0.335**
520	**0.316**	**0.295**	**0.271**
680	**0.283**	**0.258**	**0.237**
840	**0.264**	**0.236**	**0.217**

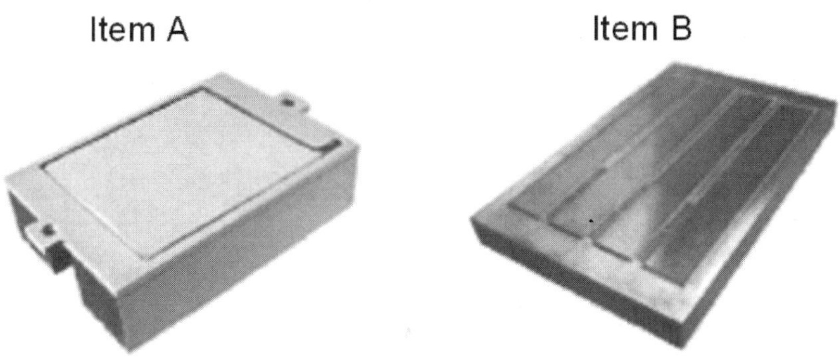

FIGURE 4.20 Various types of vapor chamber heat sinks.

face up), horizontal with top heating (fins face down), side ways, and vertical with air flow upwards. The heat sink thermal resistance which is defined as (Tb, max – Ta, in)/Q, of heat sinks at various orientations is presented in Figure 4.21.

It is a well-known fact that the surface characteristics of a heater have great effects on heat transfer in two-phase flow, especially in pool boiling. You, Rainey and Ammerman [14] have developed and patented microporous coating surface to enhance pool and flow boiling heat transfer. Based on this patent, a liquid chamber heat sink is developed. The major differences between the vapor chamber and liquid chamber are listed in Table 4.5.

4. Embedded Graphite Base
   The thermal conductivity of the natural graphite is about 200 to 400 W/m °C. However, the thermal conductivity can be increased over 1500 W/m °C after

Table 4.4 Tested Heat Sink Configurations

Manufacturers	Weight (g)	Base thickness (mm)	Descriptions
A-1	270	3	Extruded heat sink with 2 mm thick embedded sintered power VC strip in base
B-1	372	3	Wire-mesh VC base with Al zipper fins
C-1	318	3	VC base wick structure: heating side sintered powder, cooling side-wire mesh: Al zipper fins
C-2	246	3	Al base with 3 embedded heat pipes in parallel, inserted 1″ × 1″ Cu block in center of base, Al zipper fins
D-2	320	4	3 heat pipes cantilevered off Cu center block base with Al stack fins

the special heat treatment [15]. Table 4.6 [15] lists the comparisons of TPG (trademark of Momentive Performance Materials, Inc.) to other advanced thermal materials. Due to relatively high planar thermal conductivity of graphite, it has been considered as a thermal spreader. Therefore, graphite can be embedded into the heat sink base to reduce the thermal spreading resistance.

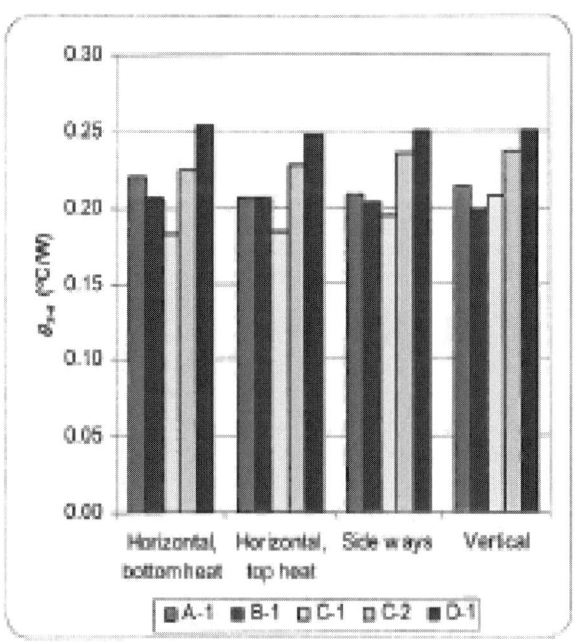

Figure 4.21 Measured heat sink thermal resistance.

Table 4.5 Tested Heat Sink Configurations

Vapor chamber	Liquid chamber
1. Two phase based on evaporation	Two phase based on nucleate boiling
2. Heat transfer by latent heat	Heat transfer including latent heat and microconvection
3. Liquid return by capillary force	Liquid return due to condensation (no wick needed)
4. Wick increasing thickness and weight	Wickless reducing thickness and weight

Table 4.6 Comparison of Various Advance Thermal Materials

Material	In-plane TC (W/m – k)	Thru-plane TC (W/m – k)	In-plane CTE (ppm/°C)	Specific gravity	Specific in-plane TC[1]
Aluminum	218	218	23	2.7	81
Copper	400	400	17	8.9	45
AlSiC-12	180	180	11	2.9	62
CuW	185	185	8.3	15.2	12
Carbon/carbon	400	40	–1.0	1.9	210
CVD diamond	1100–1800	1100–1800	1–2	3.5	310–510
TPG graphite	1500+	10	–1	2.3	650

4.2.4 Materials and Manufacture Processes

The materials for the heat sinks are generally with high thermal conductivity such as aluminum alloys (k < 240 W/m °C) and copper (pure copper k = 385 W/m °C). The former is much light and less expensive than the latter. The coefficients of thermal expansion for aluminum and copper are about 24 and 16, respectively.

The typical manufacturing processes include but are not limited to Stamping, Extrusion, Bonded/brazed, Die Casting, Forging, and Friction Stir Welding (FSW). The extruded heat sink is most popular because it is inexpensive and easy to make. However, the fin of this type heat sink is relatively thick which results in the small surface area available for the convective heat transfer due to limited number of fins on the heat sink. In addition, there is limitation for the fin height-to-fin spacing. On the other hand, there is no limitation on the fin height-to-fin spacing as well as the fin thickness for bonded fins. The typical example of this type heat sink is the plain fins or folded fins bonded to a base plate.

McLean [11] compares the heat sink thermal performance of various manufacturing processes. She first pointed out that the material thermal conductivity as listed in Table 4.7 will be increased after annealing in the heat treatment. The increased thermal conductivity is due to the improvement of the material crystal structure. As given in the same table, the material thermal conductivity with the forging process is better than those of the extrusion. The microstructures of the parts for various manufacturing processes are summarized in Table 4.8.

Table 4.7 Material Thermal Conductivity Under Various Manufacturing Processes

Raw material	Test appearance	W/o annealing thermal conductivity (W/m · K)	With annealing thermal conductivity (W/m · K)	Improvement
Copper		308.4	351.4	12.2%

Raw Material	Test appearance	With extrusion thermal conductivity (W/m · K)	With forging thermal conductivity (W/m · K)	Improvement
AL 6063	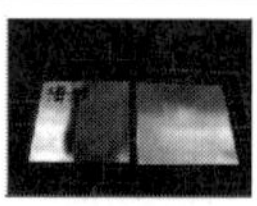	177.9	203.3	14%
AL 1070	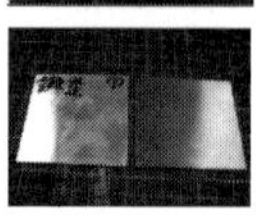	203.3	235.5	16%

Table 4.8 Microstructure of Parts (Made By Different Processes at 400×)

	Material density	Material crystal structure
Die casting 	Poor	A lot of a voids
Material injection 	Worse	A lot of void layer
Extrusion 	Fair	Particle uniform but too big
Forge 	Good	Particle very small and uniform

REFERENCES

1. Starner, K. E., and McManus, H. N., "An Experimental Investigation of Free Convection Heat Transfer from Rectangular Fin Arrays", J Heat Transfer 85, 1963.
2. Welling, J. R. and Wooldridge, C. R., "Free Convection Heat Transfer Coefficients from Rectangular Vertical Fins", J Heat Transfer 87, 1965.
3. Izume, K., and Nakamura, H., "Heat Transfer by Convection on Heated surface with Parallel Fins", Jap. Soc. Mech. Eng., 34, 1969.
4. Van De Pol, D. W., and Tierney, J. K., "Free Convection Nusselt Number for Vertical U-Shaped Channels", J Heat Transfer, 95, 1973.
5. Yeh, L. T., Yeh, Joseph and Chung, B. T. F., "Natural Convection from Finned Heat Sinks", IPack2007-33036, Vancouver, BC, Canada, July 8–12, 2007.
6. Yeh, L. T., "Optimizing Heat Sink on Chips for Natural Convection", to be published.
7. Lee, S., Song, S., Au, V., and Moran, K. P., "Closed Form Equation for Thermal Constriction/Spreading Resistances with Variable Resistance Boundary Condition", Proceedings of the 1994 IEPS Conference, 1994.
8. Song, S., Lee, S., Au, V., and Moran, K. P., "Constriction/Spreading Resistance Model for Electronic Packaging", Proceedings of the 4th ASME/JSME Thermal Engineering Joint Conference, Vol. 4, 1995.
9. Lee, S., "Calculating Spreading Resistance in Heat Sinks", Electronics Cooling, Vol. 4, No. 1, 1988.

10a. Yeh, L. T., "Thermal Characterization of Various Heat Sinks with/without Flow By-Pass", The 20th International Symposium on Transport Phenomena, Victoria, BC, Canada, July 7–10, 2009.

10b. Yeh, L. T., "Thermal Performance Evaluations of Various Heat Sinks for Air Cooling", Itherm 2815, Itherm 2012, San Diego, CA May 30 – June 2, 2012.

10c. Yeh, L. T., "Thermal Performance of Various Heat Sinks", IMPACT 2012, Taipei, Taiwan, October 24–26, 2012.

11. McLean, J. H., "Comparison of Thermal Performance for Various Thermally Enhanced Heat Sinks and Manufacturing Processes", Workshop on Thermal Management in Telecommunication Systems and Data Centers, Richardson, Texas, October 25–26, 2010.
12. Meyer, G. A., "Integrating Vapor Chambers into Thermal Solutions", Electronics Cooling, Volume 16, No. 1, 2010.
13. Glover, G., Chen, Y., Luo, A., and Chu, H., "Thin Vapor Chamber Heat Sink and Embedded Heat Pipe Heat Sink Performance Evaluations", 25th IEEE Semi-Therm Symposium, 2009.
14. You, S. M., Rainey, K. N., and Ammerman, C. N., "A New Microporous Surface Coating for Enhancement of Pool and Flow Boiling Heat Transfer", Advance in Heat Transfer, Vol. 38, 2004.
15. Liu, X., "TPG and TC1050 Materials for Thermal Management of Electronics", Workshop on Thermal Management in Telecommunication Systems and Data Centers, Richardson, Texas, October 25–26, 2010.
16. Yeh, L. T., and Chu, R. C., "Thermal Management of Microelectronic Equipment", ASME Press, 2002.

Chapter 5

SYSTEM THERMAL ANALYSIS — RACK/CABINET (PART I)

The telecommunication equipment can be deplored in the indoor or the outdoor environments, depending on its applications. Based on the functionality and performance, the indoor systems include, but not limited to, the router, the optical network, the metro-ethernet, the access, and the core network. All of these systems have the same basic mechanical configuration even though the system application areas are totally different.

5.1 KEY ELEMENTS IN SYSTEMS

A typical telecommunication rack and the side view of a shelf are given in Figure 5.1. As illustrated in the figure, the key elements include the card cage which hosts the printed circuit boards, air filter, electric magnetic insulation (EMI) plates which prevent electric magnetic waves from leaking into the ambient, and the fan trays in the forced air cooling systems. In addition, the system also includes an air inlet section and an outlet section.

5.1.1 EMI Plates

The EMI plates are generally installed on the top and the bottom of the card cage to prevent the electric magnetic leakage from the microelectronics and the power supplies to the environment. The EMI plate is a thin perforating plate (about 1 mm thick). For an effective electric magnetic insulation, the maximum percentage opening of the EMI plate is generally less than 65%.

The pressure drop can generally be expressed in the following form

$$\Delta P = K(0.5\rho V^2) \tag{5.1}$$

where K is the loss coefficient.

At the standard conditions, the pressure drop in an air flow system as presented in Equation (5.1) can further be reduced as follows:

$$\Delta P = K(V/4005)^2 \tag{5.2}$$

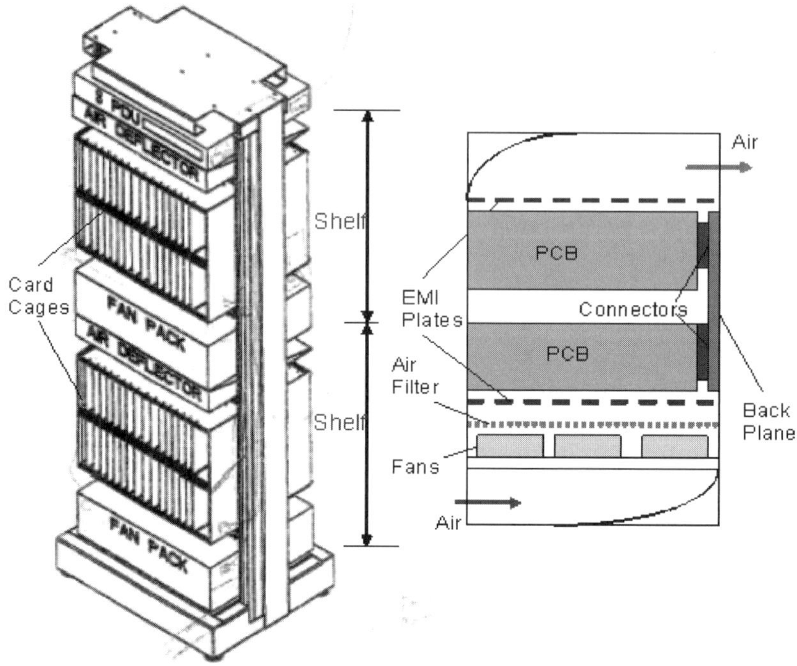

FIGURE 5.1 A typical telecom rack and sketch of shelf.

where ΔP is in inches of water and V is ft/min.

The loss coefficient across over a thin perforating plate can be expressed as follows [1]:

$$K\ (\text{loss coefficient}) = [0.707(1-f)0.375 + (1-f)]^2/f^2 \tag{5.3}$$

where f is the fraction of the total opening of a thin plate.

The loss coefficient versus the percentage of the opening is given in Figure 5.2. An additional function of the EMI (perforation) plate is to make air flow more uniform over the individual boards in the card cage due to additional flow resistance.

Recently, the honeycomb structures have been used for the EMI purpose as shown in Figure 5.3. The honeycomb structure has a much larger open than that of the perforating thin plate. The fraction of the opening for a typical honeycomb is about 0.9. However, in order to effectively reduce the electric magnetic leakage, the thickness of the honeycomb structure must be greater than 0.25 in. with preferable of 0.5 in.

Another benefit of using the honeycomb structure is to straighten the air stream as passing through the honeycomb structure. It should be noted that the pressure drop over a honeycomb structure is much higher than the value solely based on the percentage of the opening. This is due to the fact that the honeycomb structure requires a much thicker structure to prevent the electric magnetic leakage.

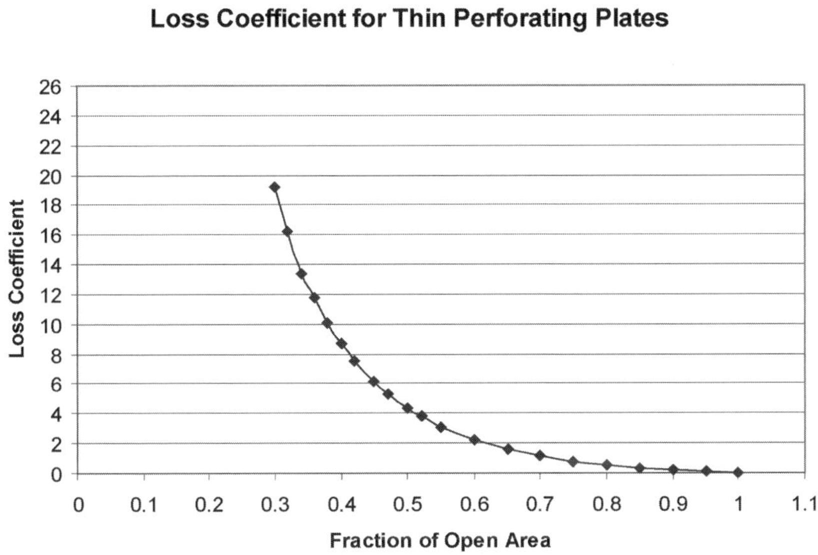

FIGURE 5.2 Loss coefficient versus fraction of opening for a thin plate.

5.1.2 Air Filters

The air filter is needed to prevent the dust or any foreign particles flowing over the microelectronics on the boards. Therefore, the filter should be placed in the upstream of the card cage. Another reason for the air filter is to create a laminar flow. The typical performance curve of a clean air filter, as shown in Figure 5.4a, is

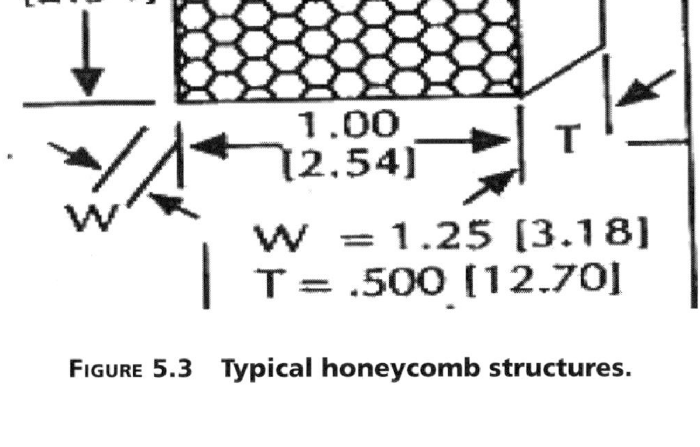

FIGURE 5.3 Typical honeycomb structures.

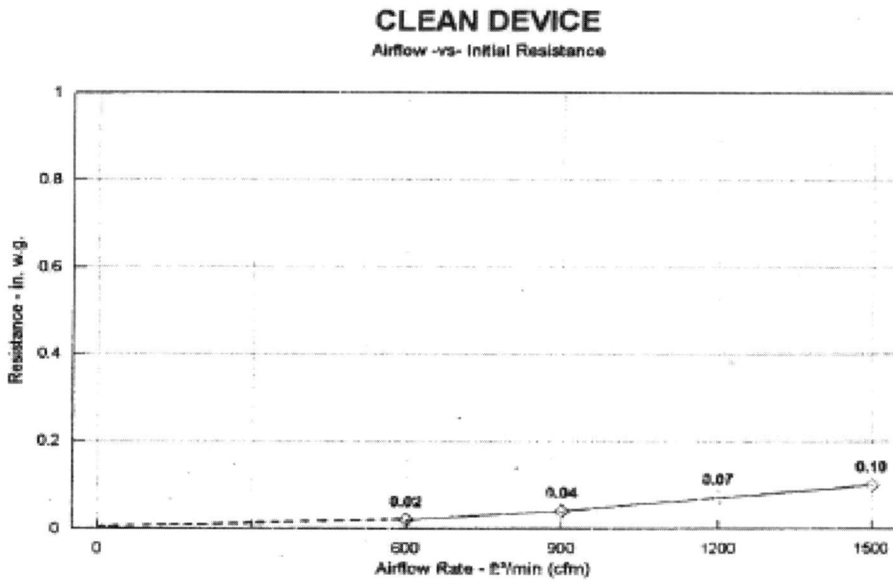

FIGURE 5.4A Performance curve of a clean air filter.

generated through a series of tests according to ASHRAE Standard 52.1-1992. For convenience, Figure 5.4a is converted to Figure 5.4b which states that the pressure loss across an air filter is a function of approaching air velocity. The pressure drop over a dirty air filter could be increased several times larger than the value of a clean one. The reason is that the majority of dusty or any foreign particles are collected on the front surface of the conventional air filters which results in a significant blockage to the air flow and leads to the high-pressure drop.

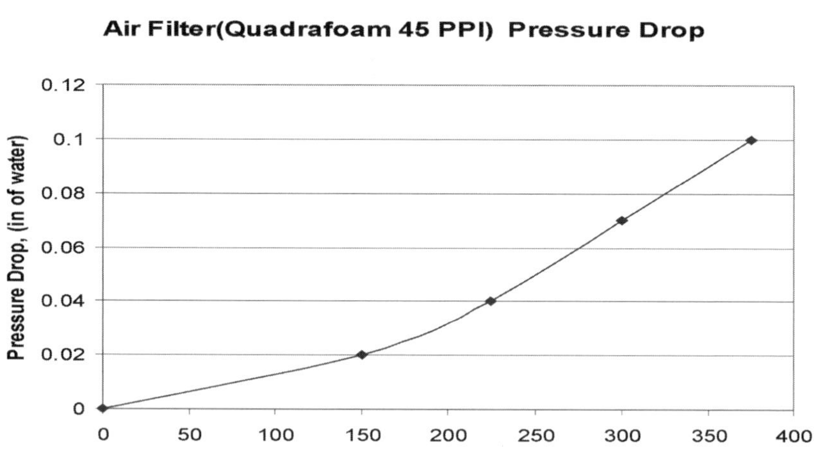

FIGURE 5.4B Pressure drop versus air flow of typical air filters.

The pressure drop across the air filter is relatively high and is a significant portion of the total pressure drop in the system. As can be seen from the above figure, the pressure drop is 0.07 in. of water at the air velocity of 300 ft/min which results in the loss coefficient of the filter (K) is 12.5 according to Equation (5.2). On the contrary, the loss coefficient for a typical board (not very high dense board) is probably less than 6 or 7.

Consideration must be taken to reduce the pressure drop over the filter. Since $\Delta P \, \alpha V^2$, Equation (5.1) can be rewritten as follows:

$$\Delta P \, \alpha (1/A)^2 \tag{5.4}$$

where A is the cross-section area of the filter.

Therefore, one way to reduce the pressure drop is to have as large air filter as possible. The best location is to place the air filter under the card cage as shown in Figure 5.1 because this position generally gives the largest surface area (over the width and the depth of the rack). The air filter can be either placed in the upstream or downstream of the fans. The former will prevent dust over the fans, and the latter will make air flow more uniformly flow over the individual boards due to the high flow resistance of the air filter.

5.1.3 Fans

The fan is the heart of any forced air cooling system. The fan propeller converts torque from the motor to increase the static pressure cross the rotor of the fan and to energize the air particles. The discussion here is focusing at the fan application in the rack configuration.

Thermally, the location of the fans in a system is also an important factor to consider. The fan located at the exit of a system is usually referred to as an exhaust fan which draws air through the system. This type of installation reduces the internal system pressure (to less than the ambient pressure) and possibly draws dust or foreign particles into the system. The advantage of this method is that no additional heat dissipated from the fans is added to the system. However, the fans operate at higher air temperature at the exit and result in reduction of the fan reliability as well as the fan life.

On the other hand, a blowing fan, which is located at the inlet or upstream of the card cage in the system, creates a higher internal pressure and keeps dust or any foreign particles out of the system. However, the inlet air is heated up by the heat dissipation from the fan and thus the air temperature is increased prior to entering the system or the card cage where the printed circuit boards are located.

Typically, multiple fans which are arranged in parallel are placed in a fan tray. Therefore, the acoustic noise of the system is significantly increased with the number of fans. The accumulated delta noise level can approximately be determined by the following equation:

$$\Delta(\text{dBA}) = 10 \log (N) \tag{5.5}$$

and, the total system acoustic noise is given as follows

$$(\text{dBA})\text{system} = (\text{dBA})\text{sf} + \Delta(\text{dBA}) \tag{5.6}$$

where N is the number of fans in the system and (dBA)sf is the acoustic noise of a single fan.

5.2 SYSTEM LEVEL THERMAL ANALYSIS

The ultimate goal of the thermal analysis is to maintain the temperature of all electronic components in the system below their required limit. The system level thermal analysis is first performed to calculate system flow rate as well as the flow rate to individual boards which is often also referred to as the slot flow rate. The computed slot flow rate is then applied to the individual boards for the board level thermal analysis which is to determine the temperature of the components on the individual boards.

Since the detailed information of individual boards is generally not available in the early design phase, one must assume some values of pressure drop for the boards. This assumed value can be obtained from the existing similar boards or just from the preliminary board layout. The assumption of this value is not very critical for the purpose of the system level analysis. The reason is that the pressure drop over the board is typically less than one third of the total system pressure drop. Therefore, the effects of the errors in the estimated pressure drop over the board on the system flow rate will be limited. In addition, the more accurate estimated pressure drop over the board will be available once the detailed board is developed.

No detail of the board layout is needed for the system level analysis. Each board can be represented by a large component on the board with the board heat dissipation. The purpose of this assumption is to provide correct heat dissipation from individual boards in order to compute air temperature rise over the boards and also to show the component side of the boards. Furthermore, the assumed loss coefficient over individual boards is then treated as the flow resistances at the inlet and the exit of the boards. Another popular way to predict the air temperature rise is to treat the power from the board as an internally volumetric heat generation in the space, and the volumetric flow resistance is assumed over this space.

Figure 5.5 shows the simple sketch of a representative telecommunication rack and its flow resistance over the system. The individual flow resistance is as follows:

R1 = entrance
R2 = inlet section
R3 = 90° turn
R4 = air filter
R5 = card cage, including 2 EMI plates
R6 = 90° turn
R7 = exit section
R8 = exit

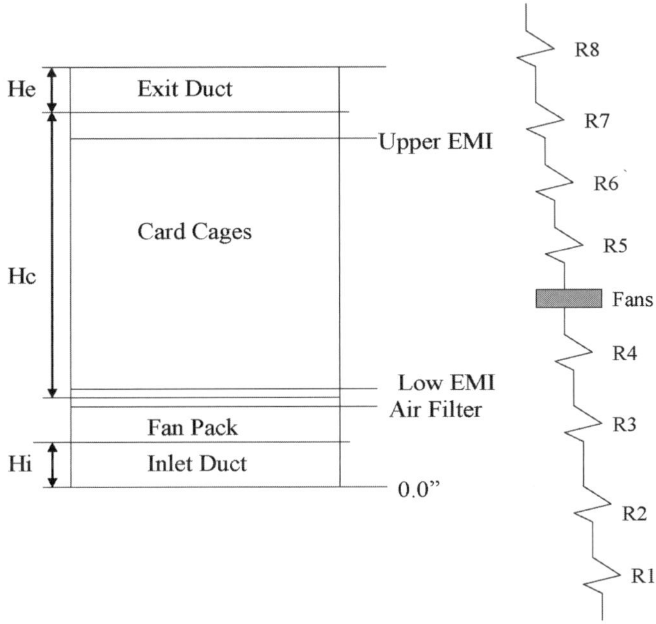

Figure 5.5 Sketch of typical rack and its flow resistance.

The flow resistance can be derived from Equation (5.1)

$$\Delta P = K\,(0.5\rho V^2) = RQ^2 \tag{5.7}$$

where R is the flow resistance, and Q is the volumetric flow rate.

The intersection of the fan performance and the system resistance curve is the system operation point, and the flow rate at the operation point is the average system flow rate. In theory, the estimated system flow rate can be obtained by the hand calculations. However, a CFD (computational fluid dynamics) analysis with aids of the commercial software is typically employed in determining the system flow rate as well as the flow rate to individual boards.

As stated previously, the system flow rate as well as the board flow rates can be determined by the hand calculations instead of using the CFD. The hand calculation method is in general based on the flow resistance network. According to the flow network modeling (FNM), the software analysis tool, "MacroFlow" [2] was developed by Innovative Research for the system level flow analysis.

The CFD analysis is used to compute the velocity and pressure of each fluid particle, while the FNM analysis is based on the lumped system and determines the average value of the velocity and pressure of the system. The FNM method is extremely efficient and fast for the system level flow analysis. For example, it generally takes 3 to 5 days to build a system level CFD model of a typical telecommunication rack and may take several hours running the simulation to obtain the converged results. On the other hand, it takes a couple of hours to develop FNM model and less than 30 sec to obtain the solution. In most cases, the difference between the FNM and the CFD results is generally less than 10%. The FNM is very useful tool in the early phase of the system design, especially in defining the system concepts. However, it cannot be adapted for the detailed thermal design of the system.

5.3 PRACTICAL EXAMPLES

Example 5.1

A telecommunication rack consists of two identical and independent shelves as shown in Figure 5.6 [3]. For clarity, the front door is not shown in the figure. Each

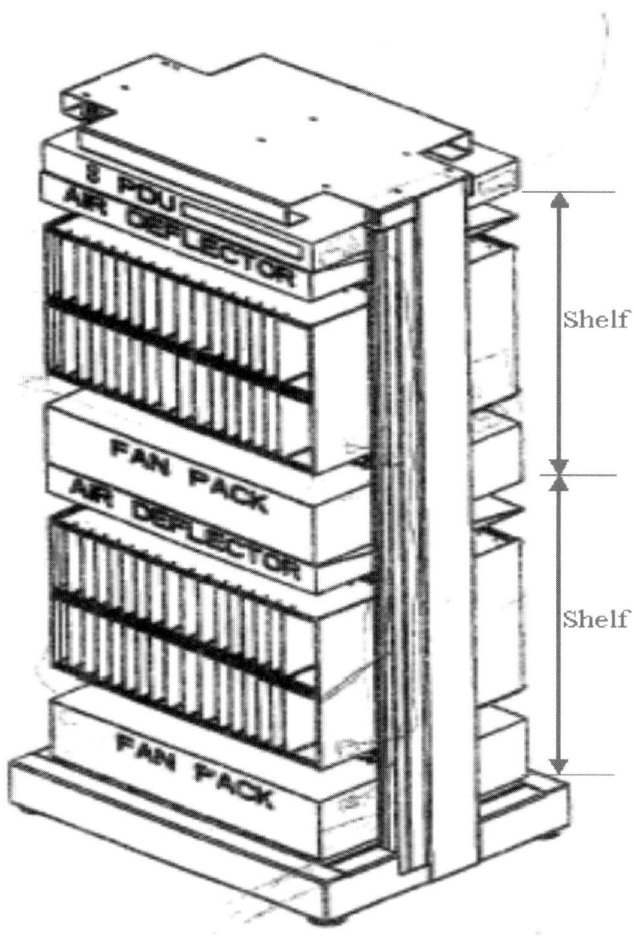

FIGURE 5.6 Telecommunication rack with two identical shelves.

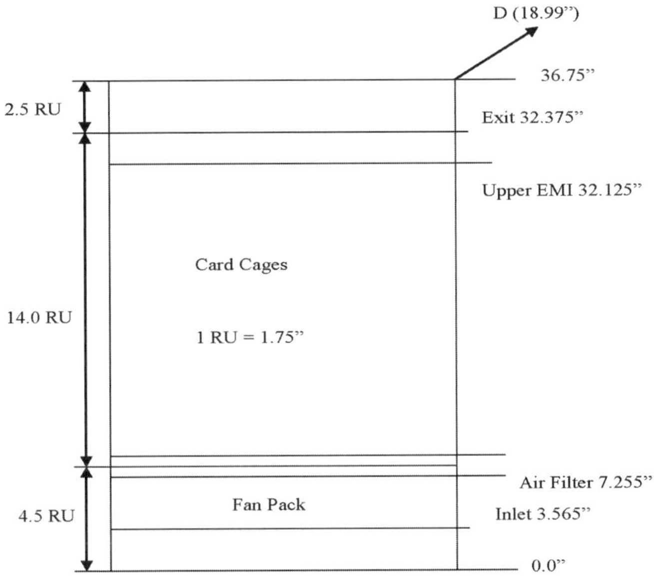

FIGURE 5.7 Sketch of shelf.

rack consists of two identical shelves located on the top-to-bottom orientation. Each shelf with overall dimensions of 25.45 in. (width) × 18.99 in. (depth) × 37.75 in. (height) includes one fan tray with six fans, and three card cages with a total of 50 printed circuit boards (PCBs). The sketch of the shelf is given in Figure 5.7. Air at 40°C enters through the door from the front of the shelves, and then makes a 90° turn upwards through PCBs, and finally turns another 90° to exit the system from the rear of the shelf. The system level analysis is performed independently on each shelf.

The only difference between the upper and lower shelves in the rack is at the opening of the front door. The shape and size of the opening (or air inlet) on the front door are different. The lower shelf has a rectangular opening with the cross-section area (90.73 in.^{2}) identical to the inlet of the shelf. On the other hand, the door opening at the upper shelf has an odd shape as shown in Figure 5.8 with the cross-section area of 70.38 in^2 which is smaller than the inlet of the shelf. The odd shaped opening at the front is approximated by four different sizes of the rectangular openings in the thermal model as presented in Figure 5.9. The total cross-section area of these four rectangular openings is equal to that of the odd shaped opening. The major components, such as the fan tray and three card cages inside the shelf, are included in the detailed thermal mode as shown in Figure 5.9.

The bottom of each shelf is a fan tray with the height of 7.875 in. The fan tray includes 6 EBM (DV6248TD) fans with each one consuming 90 W and generating acoustic noise of 69 dBA. The outer diameter and the thickness of the fan are 6.75 in. and 2 in., respectively. The fan operates at 48 V in the normal operation The fan locations relative to the mid-plane are shown in Figure 5.10.

The shelf consists of three card cages with the total power of 5626 W that are located above the fan tray. The shelf is subdivided into the front and the rear sections

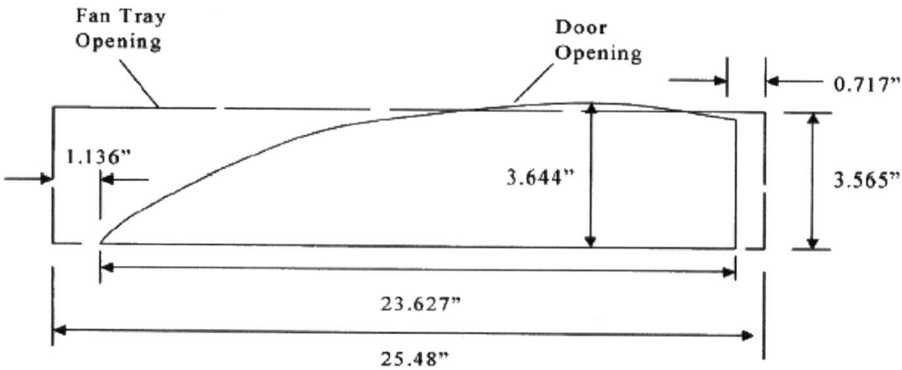

FIGURE 5.8 Odd-shaped opening at upper section of front door.

by a mid-plane. The front section of the shelf includes two identical card cages with the total power of 2246 W, while there is only one card cage in the rear section with the power of 3380 W. The mid-plane provides all electrical connection to the boards inside the card cages. The air exit is located on the top of the card cages. In addition, there are two perforate plates in the shelf for EMI (electrical magnetic insulation) protection purpose. For simplicity, the air filter and the EMI plates are modeled as a planar flow resistance with a zero thickness. The loss coefficients for the EMI plates and the air filter are 1.7 and 14, respectively. The loss coefficient

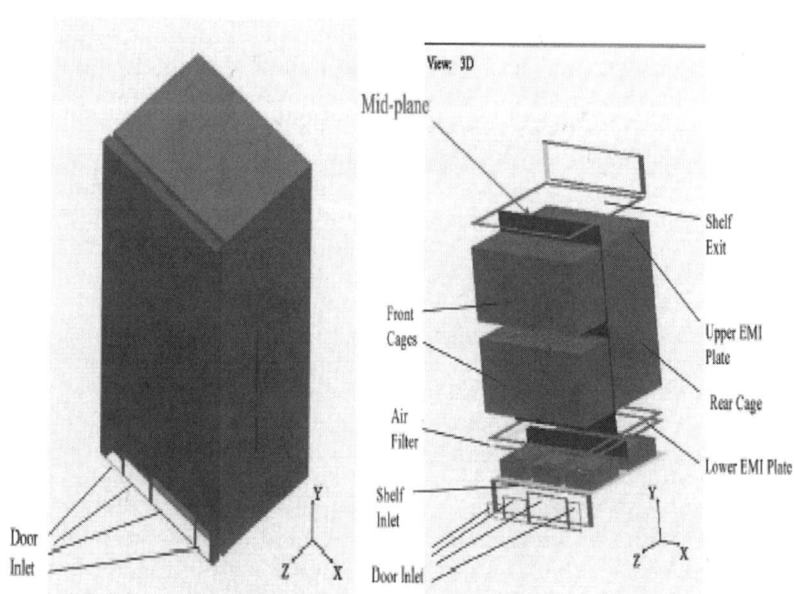

FIGURE 5.9 System thermal model.

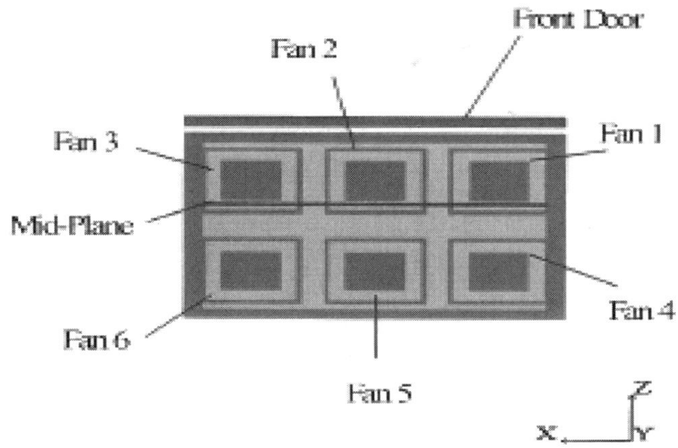

FIGURE 5.10 Fans locations.

of 1.7 corresponds to 63% opening on the thin perforation plate. In addition, the ambient condition is assumed to be at 40°C and at the sea level.

The main purpose of the CFD analysis is to determine the air flow rate to individual boards. The predicted air flow rates are then applied to the detailed board level thermal analysis. The component junction temperatures and the pressure drop across the boards are computed from the board level analysis of individual boards. The total number of the cells in the current model is over 433,300. There is no need to discuss the details of fundamental heat transfer theories involved here because of using a commercial software tool.

The card cage sketch and the flow rate to individual boards for the upper front cage are presented in Figure 5.11a. The GR-63 also requires the system to meet the

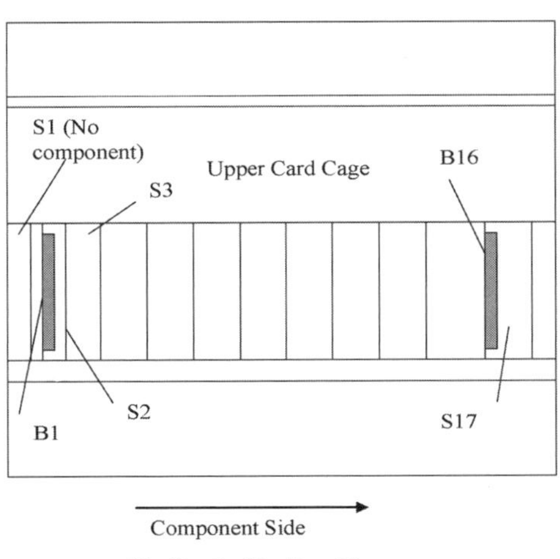

Slot	CFM	Slot	CFM
1	9.6 (8.7)*	2	31.4 (27.1)
3	24.6 (20.7)	4	24.4 (20.2)
5	23.8 (18.8)	6	23.7 (16.9)
7	24.0 (15.9)	8	30.8 (20.2)
9	38.9 (22.7)	10	31.7 (17.1)
11	31.7 (15.6)	12	24.8 (10.9)
13	24.9 (11.4)	14	25.5 (12.3)
15	26.1 (12.6)	16	26.1 (13.8)
17	32.9 (19.7)		

* ( ) Fan 3 Failure

Total flow : 455 CFM (285 CFM)

FIGURE 5.11A Card cage sketch and board flow rates for front card cage.

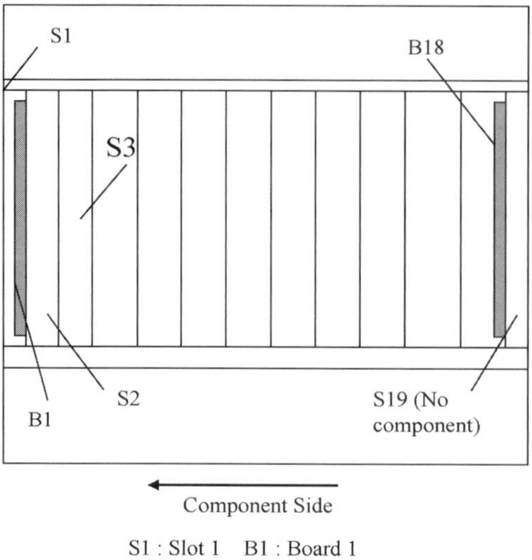

Slot	CFM	Slot	CFM
1	35.0	2	67.4
3	55.5	4	50.8
5	51.7	6	51.1
7	52.7	8	66.0
9	67.7	10	53.8
11	66.4	12	52.7
13	52.5	14	51.8
15	51.9	16	52.2
17	64.4	18	68.9
19	7.5		

Total flow : 1020 CFM

FIGURE 5.11B Card cage sketch and board flow rates for rear card cage.

thermal requirements under one fan failure condition. The flow rate for individual boards for the case with Fan 3 failure is also included in the figure. The analysis should be repeated for every fan failure condition so that the worst case can be identified. The board flow rates for the rear card cage are presented in Figure 5.11b.

As can be seen from the above figures, the flow rates to the front and the rear card cages are 455 and 1020 CFM, respectively. The total flow rate for the upper shelf is 1475 CFM with the system pressure drop of about 1.1 in. of water. It should be noted that the operation point of the individual fans varies slightly even for the identical fans. This is mainly due to the fan locations. Examining the results in details also reveals that 69% of the total system flow goes to the rear card cage which has 60% of the total system power.

Example 5.2

Another example is to employ a CFD analysis to examine the effects of the air filter locations at the upstream and downstream of the fans on the flow distribution within the system [4]. A telecommunication shelf as shown in Figure 5.12 includes a card cage, a fan tray, an inlet, and an exit. The system includes four EBM (DV6248TD) fans as adapted in Example 1. The overall internal dimensions of the shelf are 21.2 in. (width) × 36.75 in. (height) × 19 in. (depth). There are four compartments inside the card cage. Each of three identical compartments contains an optical electronic module (OEM) and its motherboard. The fourth compartment includes four small printed circuit boards.

The present discussion is focusing at the effects of the air filter location relative to the fans on the system air flow rates. Parametric studies are performed to examine the effect of the location of the air filter on the flow distribution. Figure 5.13 shows the results for the case with four fans in operation. The difference in the flow

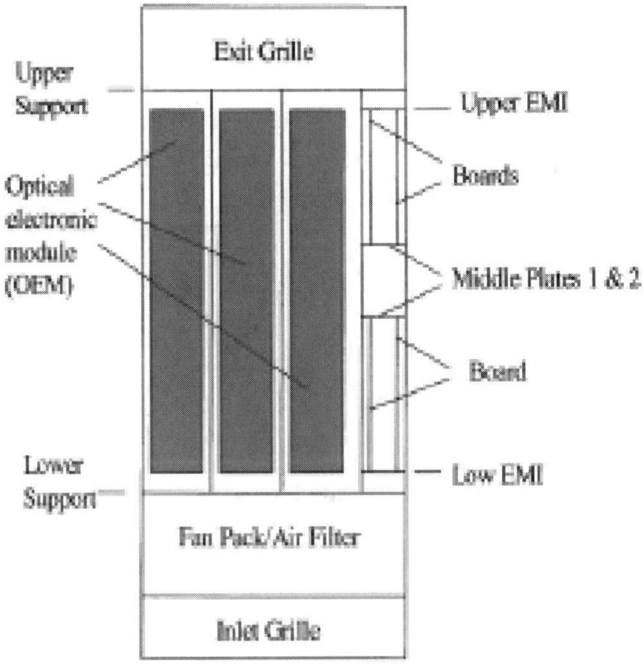

FIGURE 5.12 **Sketch of shelf under consideration.**

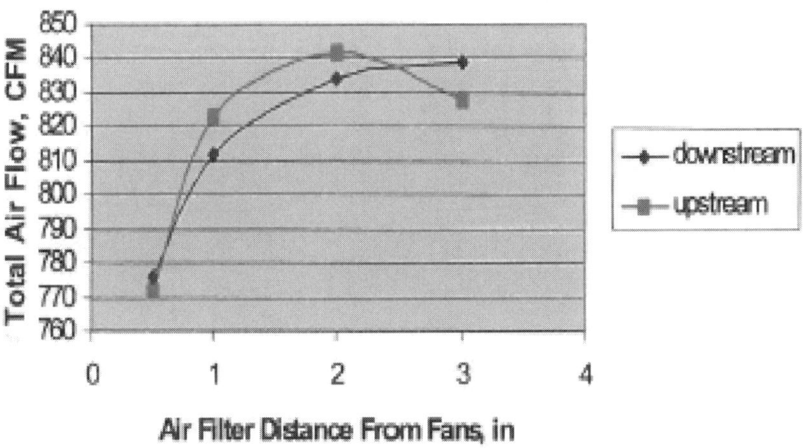

FIGURE 5.13 **Air flow rates versus filter locations with four fans in operation.**

Table 5.1 Air Flow Rate for Four Fans in Operation with Air Filter 3 in. from Fan

Compartment	Up-stream filter flow rate (CFM)	Down-stream filter flow rate (CFM)
OEM1	253	243
OEM2	208	248
OEM3	289	223
Boards	78	125
Total	828	839

rate between the cases with the air filter being located at the up- and the down-stream of the fans is small (less than 11 ft^3/min (CFM)) regardless the location of the air filter.

For down stream air filter case, the total system flow rate increases as the distance from the fan pack increases. However, on the other hand, an optimum distance of 2 in. from the fan pack that generates the maximum system flow rate is achieved for the up-stream air filter. The flow rate distribution to individual compartments with the air filter located at 3 in. from the fan is given in Table 5.1.

The results for the case with the one failure are presented in Figure 5.14. The optimum distance for both cases (up- and down-stream air filters) is found to be about 1 in. from the fan pack. Table 5.2 gives the flow rate to individual compartments for one fan failure. The failed fan is located just below OEM1 and OEM2. Therefore, little flow is supplied to these two modules when the air filter is located upstream of the fans as given in Table 5.2. The table also indicates that the filter located at the upstream of the fan provides uneven air flow to the compartments when one fan fails. On the other hand, the case with the air filter placed in the down stream of the fans gives a more uniform flow distribution to all compartments as given in Table 5.2. This is because the latter pressurizes the plenum due to a relatively high flow resistance of the air filter.

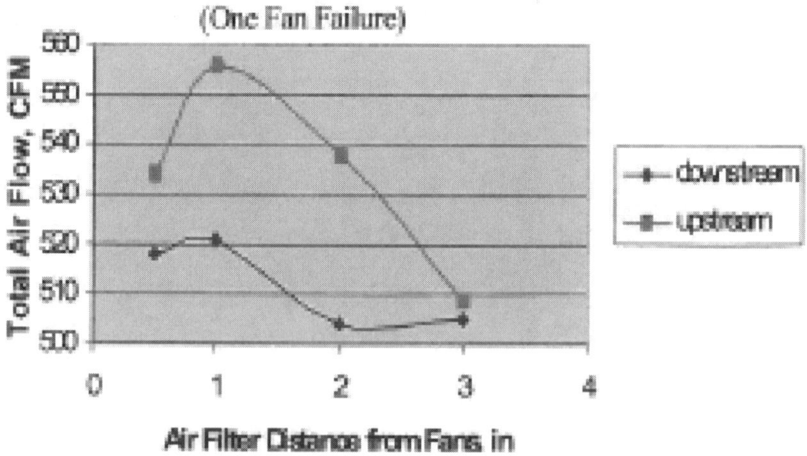

FIGURE 5.14 Air flow rates versus filter locations for one fan failure.

Table 5.2 Air Flow Rate for Three Fans in Operation with Air Filter 3 in. from Fan

Compartment	Up-stream filter flow rate (CFM)	Down-stream filter flow rate (CFM)
OEM1	37	144
OEM2	45	126
OEM3	335	142
Boards	92	92
Total	509	504
Reversed flow (leakage from bad fan)	270	321

The real concern is not at the total system flow rate but the flow rate to individual modules. As can be seen from the above table, the flow rate to OEMs 1 and 2 for the case with the up-stream air filter is much smaller than those with the downstream air filter. This may lead to an overheating of the components inside OEMs 1 and 2. The flow leakage from the failed fan for the down-stream filter is much larger than that of the upstream filter. This is because of the high flow resistance of the filter.

It is a well-known fact that the flow rate generated by a fan will be decreased if the fan is placed too close to any structures. The general rule is that the distance between the fan and the blocking structure must be at least greater than the half of the diameter of the fan. With the limited data available from this study, this general rule seems to be also valid for the array of the fans.

To eliminate or reduce the flow leakage from a failed fan, a damper as shown in Figure 5.15 may be added to each fan. However, the damper generally increases the flow resistance and thus decreases the available air flow to the system. Another concern about the damper is the possibility of failure. The fan may be blocked if the damper fails to open on the good fans. The practice of adding the fan cover

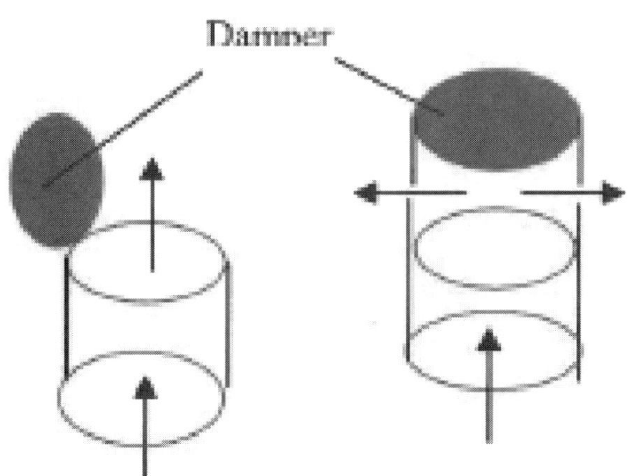

FIGURE 5.15 Examples of fan damper/cover.

has been increasingly adapted in the industry, and the trends are expected to grow in the future.

5.4 THERMAL DESIGN PROCESS

In order to insure the quality and the accuracy of the thermal analysis results, one must have a good design process to follow. The recommended design process presented below is summarized in the following design flow chart as presented in Figure 5.16.

The flow chart below describes the complete thermal design process for a telecommunication rack. The entire design process can be divided into 3 phases which are summarized in the following steps:

1. Phase 1: System level design to determine the flow rate to individual boards.
2. Phase 2: Board level design to compute the junction temperature of all components.
   (a) Using the computed the air flow rate to individual boards along with the inlet air temperature to perform the detailed analysis for all boards.
   (b) Design refinements are often required in order to meet the component temperature limits. The most common refinement is to resize the heat sink on the hot components. Others include the relocations of the hot components or adding the baffles to guide the air flow to the hot components or employing high thermal conductivity TIMs (thermal interface materials).

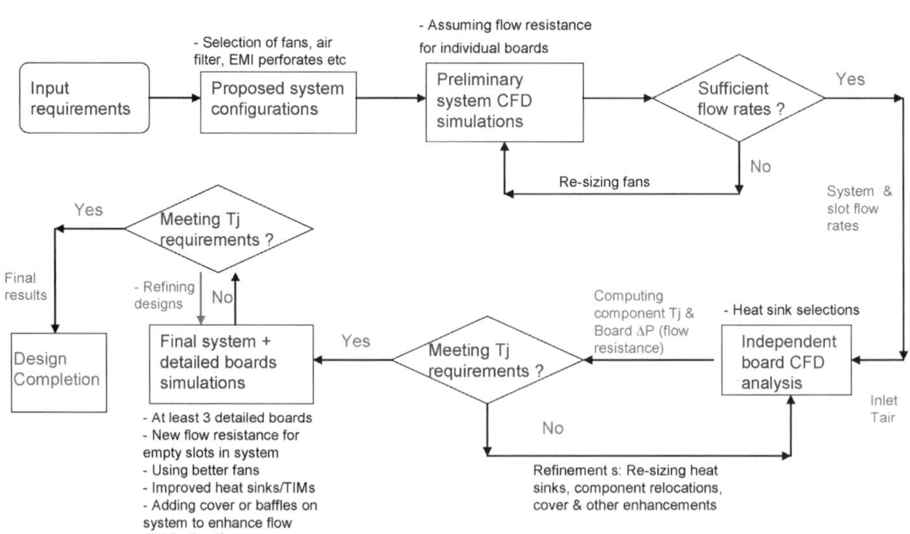

FIGURE 5.16 Telecommunication rack thermal design process.

(c) The results from the board level analysis are the component junction temperatures and the pressure drop (or the flow resistance) of the individual boards.

3. Phase 3: Final design analysis to combine system and boards, including 3 detailed boards.
   (a) The component temperature on a given board will be affected by the presence of neighboring boards. Therefore, three detailed boards are needed in the final analysis. The target board is sandwiched by two detailed boards. The pressure drop or flow resistance calculated from the board level analysis should be applied to the rest of other boards. For the board located at both ends of the card cage, two boards (the target board and the neighbor board) will be needed in the final analysis.
   (b) The component temperatures obtained for the target board are the final results.
   (c) The analysis should be repeated by rotating three-board group throughout the entire system.

Theoretically, the final thermal model should include all detailed boards in the system. Unfortunately, the thermal model in such case will be too large to run and become unpractical. The adoption of the three-board group concept is to obtain the accurate results within the reasonable time. In addition, the effects from the boards located outside the three-board group are generally very limited.

As required by the GR-63, the thermal analysis must be performed for the worst case of one fan failure condition. In theory, one must turn off the fan one by one to determine the worst case of one fan failure case. However, most experienced thermal engineers can quickly narrow down to a couple of possible worst cases to be analyzed.

REFERENCES

1. Blevins, R. D., Applied Fluid Dynamics Handbook, Van Nostrand Reinhold, New York, 1984.
2. Macroflow, a product of Innovative Research, Inc, Minnesota.
3. Yeh, L. T., "System Level Thermal Analysis of Large Telecommunication Racks", InterPack2003-35047, InterPack Conference, 2003.
4. Yeh, L. T., Chung, B. T. F., and Yeh, J., "A CFD Analysis of Telecommunication Racks, including Effects of Air Filter Locations", InterPack2005-730008, InterPACK Conference, 2005.
5. Yeh, L. T., and Chu, R. C., "Thermal Management of Microelectronic Equipment: Heat Transfer Theory, Analysis Methods and Design Practices", ASME Press, New York, 2002.

Chapter 6

SYSTEM THERMAL ANALYSIS — RACK/CABINET (PART II)

The system level thermal analysis and the complete thermal design process of a telecommunication rack have been discussed in the previous chapter. Basically, the system flow rate is determined by the fan operation point which is defined as the intersection point of the fan performance curve and the system pressure drop curve. The system pressure drop is the sum of the pressure drop over the inlet, turns, air filter, EMI plates, boards, and exit.

This chapter will discuss the effects of the system configuration on the system flow rates and various subjects related to the system thermal design.

6.1 EFFECTS OF SHELF CONFIGURATION ON SHELF FLOW

The effects of system height and/or depth on the system flow rates [1] will be discussed. As stated above, the system flow rate depends on the system pressure drop (i.e., the system flow resistance). Figure 6.1 which was reproduced from Figure 5.5 shows the simple sketch of a representative telecommunication rack and its flow resistance over the system. The individual flow resistance is as follows:

R1 = entrance
R2 = inlet section
R3 = 90 degree turn
R4 = air filter
R5 = card cage, including 2 EMI plates
R6 = 90 degree turn
R7 = exit section
R8 = exit and
Rt = total system flow resistance
= R1 + R2 + R3 + R4 + R5 + R6 + R7 + R8

The flow resistance can be expressed as follows

$$\Delta P = K(0.5\,\rho V^2) = R\,Q^2 \tag{6.1}$$

where K is the loss coefficient, R (= 0.5 $K\rho/A^2$) is the flow resistance, Q (= AV) is the volumetric flow rate, A is the cross-section area, and V is the average velocity across this cross-section area.

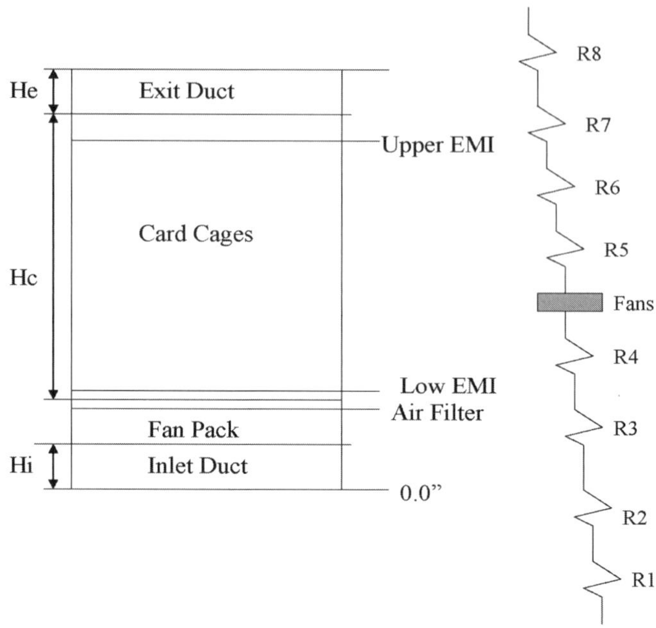

FIGURE 6.1 Sketch of typical rack and its flow resistance.

This study is to examine the effects of various parameters of the shelf configuration on the system flow rate. The results predict the changes of the flow rate over the value of the baseline configuration which is a known factor. The purpose of the analysis is to develop a family of curves serving as the design guidelines, which can quickly predict the effect of the changes of individual parameters on the system flow rate.

6.1.1 Effect of Shelf Height

Figure 6.2 illustrates the system configuration under consideration. Hi is the height of the inlet and exit. It should be noted that the height of the inlet can be different from that of the exit. For simplicity, both the inlet and exit assume to have the same height. Additional assumptions are as follows:

1. All flow resistance from R1 through R8 of the baseline configuration are known.
2. Except the height, all other dimensions of the shelf remain unchanged.
3. Flow resistance of other sections or parts of the system are unchanged.

From Equation (6.1), one obtains the following relationship for the inlet and the exit

$$R\alpha(1/H^2) \tag{6.2}$$

where H is the height of the inlet and the exit sections.

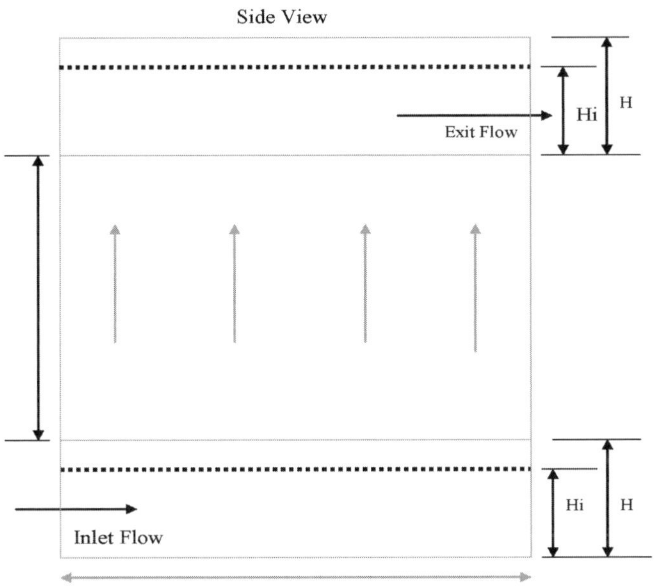

FIGURE 6.2 Sketch of shelf with various inlet/exit height.

Based on the above assumptions,

$$R/(R)\text{bl} = (\text{Hi}/H)^2 \tag{6.3}$$

where (R)bl and Hi are the flow resistance and inlet/exit height of the baseline configuration. Equation (6.3) states that increasing the inlet height (H > Hi) will decrease the inlet flow resistance. The next step is to calculate the new total system flow resistance, Rt based on the calculated flow resistance for the inlet/exit. Once the new system flow resistance is known, the system new flow can readily be determined by the following equation

$$Q/(Q)\text{bl} = [(\text{Rt})\text{bl}/\text{Rt}]^{0.5} \tag{6.4}$$

The following example is to show how to compute the new system total flow resistance and the flow rate:

Example 6.1: For *H*/Hi = 1.5,

From Equation (6.3), one obtains

$$\text{R18} = 0.4444(\text{R18})\text{bl} \tag{6.5}$$

If (R18)bl (= R1 + R8)bl = 0.1(Rt)bl for the baseline configuration, then the new total system flow resistance for H/Hi = 1.5 is

$$\text{Rt} = 0.1 \times \text{R18} + 0.9(\text{Rt}) = 0.94444(\text{Rt})\text{bl} \tag{6.6}$$

and the new volumetric flow rate with H/Hi = 1.5 is

$$Q/(Q)\text{bl} = [(\text{Rt})\text{bl}/\text{Rt}]^{0.5} = [0.94444]^{0.5} = 1.029 \quad (6.7)$$

Equation (6.7) implies that the inlet/exit height increases by 50%, the system flow rate increases by about 3% for (R18)bl = 0.1(Rt)bl if other parameters of the system remain unchanged.

Following the above procedure, the family of the curves is developed as shown in Figures 6.3 and 6.4 for the new flow resistance and the new flow rate, respectively.

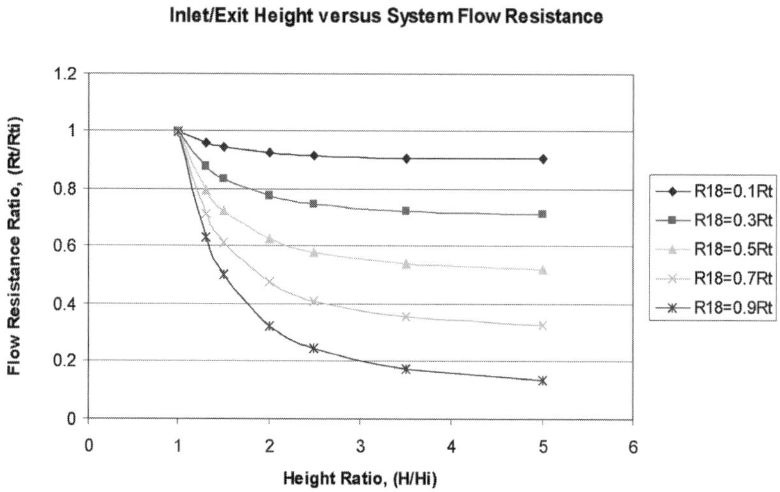

FIGURE 6.3 Inlet/exit height versus system flow resistance.

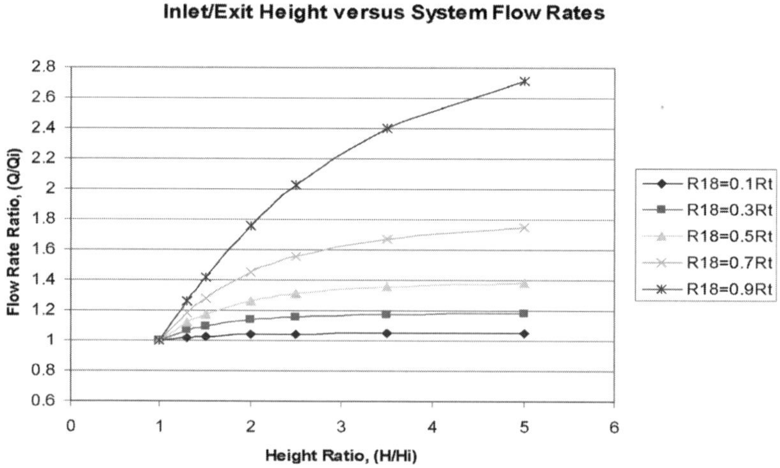

FIGURE 6.4 System flow rate ratios over inlet/exit height ratios.

6.1.2 Effect of Shelf Depth

Figure 6.5 will be used to examine the effect of the shelf depth on the system (shelf) flow rate. The analysis is based on the following assumptions:

1. All flow resistance from R1 through R8 of the baseline configuration are known.
2. Except the depth, all other dimensions of the shelf remain unchanged.
3. Flow resistance of other sections or parts of the system is unchanged.

From Equation (6.1), one obtains the following relationship for the card cage plus with two EMI plates

$$R5/(R5)bl = [(Ac)bl / Ac]^2 \qquad (6.8)$$

where Ac (= W^*D) is the cross-section area of the card cage normal to airflow direction which is identical to the cross-section area of the air filter. W is the width of the shelf and D is the depth of the shelf.

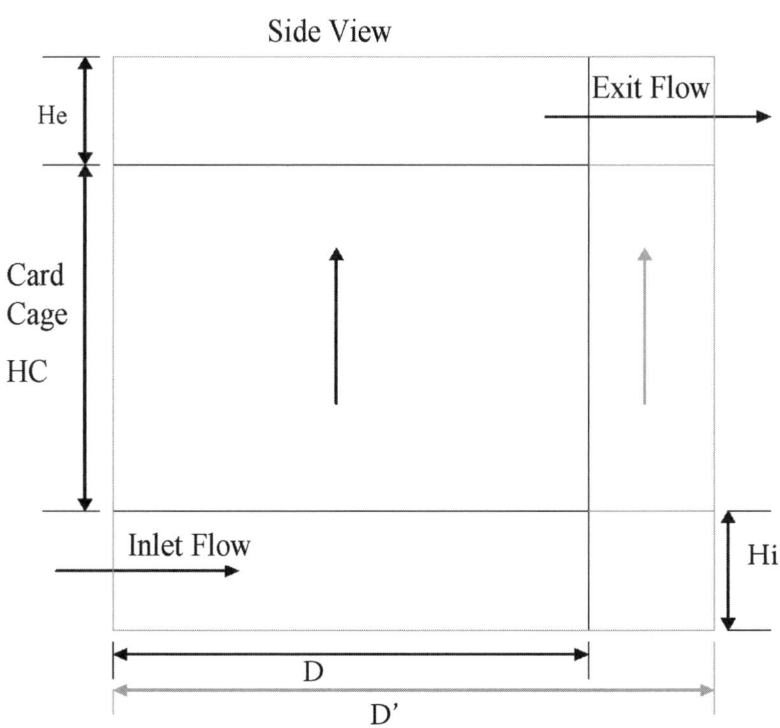

FIGURE 6.5 Sketch of shelf with various depth.

Equation (6.8) is further reduced to

$$R\alpha(1/D^2) \tag{6.9}$$

Similar to the above calculating procedure, the next step is to compute the new system flow resistance and finally the new system flow rate. The effects of the shelf depth on the flow resistance and the flow rate are given in Figures 6.6 and 6.7.

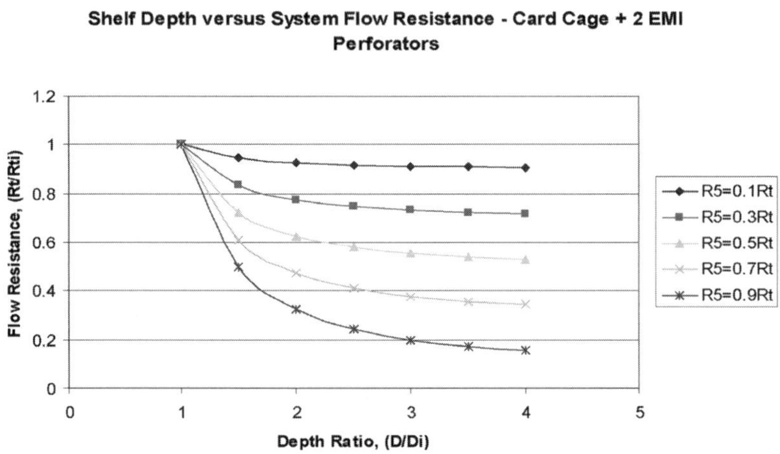

FIGURE 6.6 Shelf depth versus shelf flow resistance.

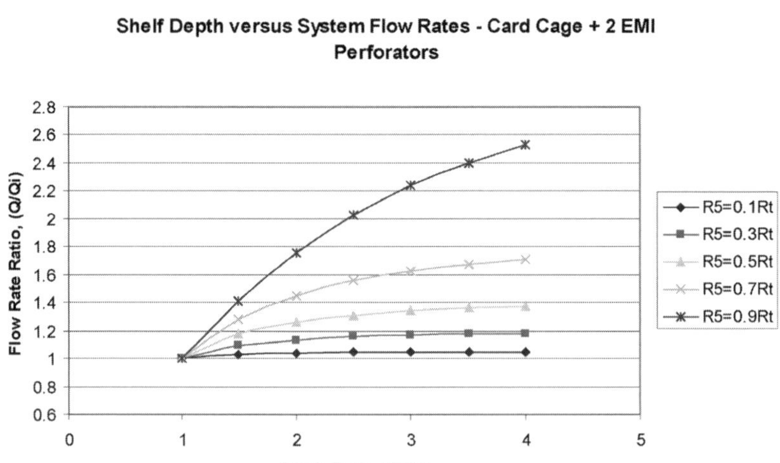

FIGURE 6.7 System flow rate ratios over shelf depth ratios.

6.1.3 Design Optimization Procedure

For convenience to users, the new family of curves is developed based on R1278 and R45. The R1278 represents the combined flow resistance of the entire section of the inlet and the exit. Similarly, the R45 corresponds to the combined flow resistance of the card cage (over printed circuit boards), two EMI plates and the air filter. The new family curves are presented in Figures 6.8 and 6.9 for the effects due to varying height and depth, respectively.

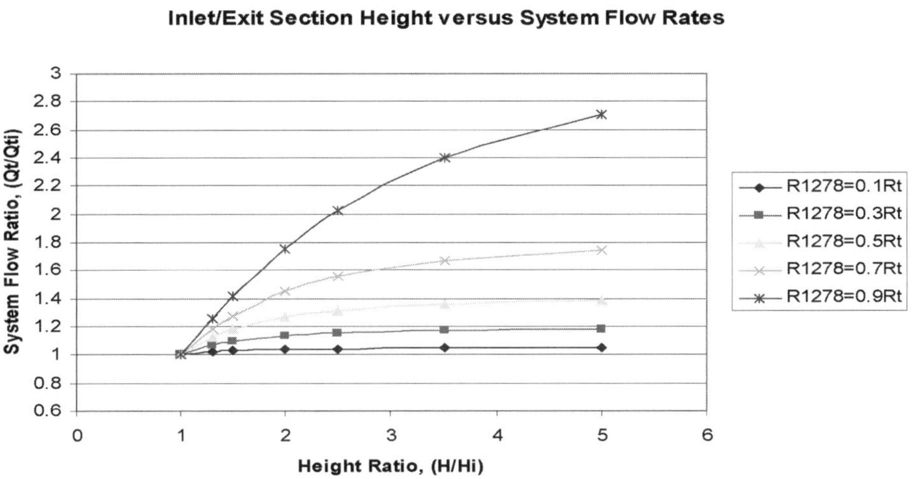

FIGURE 6.8 Effect of inlet/exit section height on system flow rate.

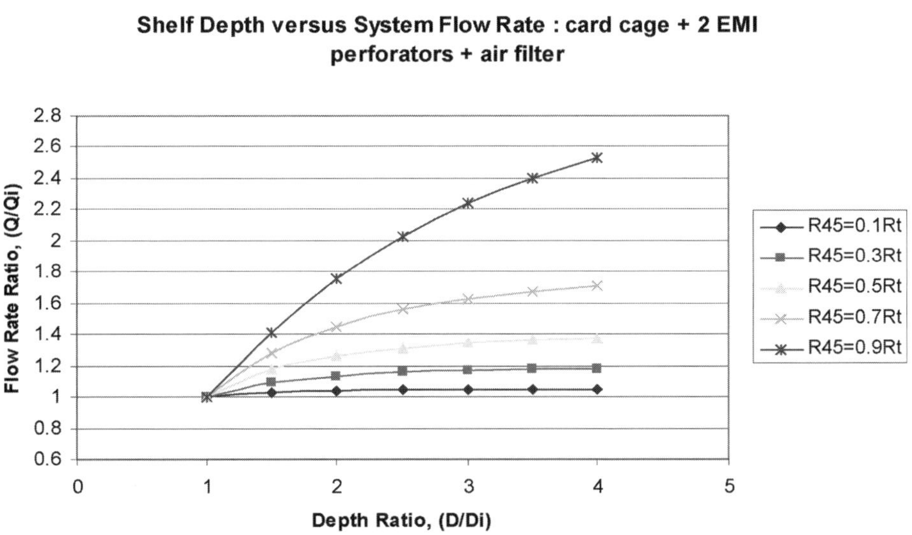

FIGURE 6.9 Effect of depth on system flow rate.

The design optimization procedure can be described as follows:

1. Performing CFD analysis or testing on the initial baseline configuration to obtain the system flow rate, Q0 and all flow resistance (R1 through R8)bl. The obtained results will be the starting point in the above figures.
2. Increasing the height of the inlet and/or exit section to a desired value while keeping the shelf depth unchanged. The new system flow rate, Q1′ (= Q1/Q0) (dimensionless) at the adjusted height can be computed from Figure 6.8, and this new value will serve as the starting point of Figure 6.9.
3. Varying the depth to the desired value with the height of the inlet and exit section fixed at the value of Step 2. The system flow rate, Q2′ (= Q2/Q1) (dimensionless) for the adjusted depth can be determined from Figure 6.9.
4. The system final dimensionless flow rate is the product of Q1 and Q2 and the dimensional flow rate is $Q0 \times Q1' \times Q2'$.

It should be noted that the sequence of Steps 2 and 3 can be interchangeable. The present analysis is to provide a quick result to evaluate the effects of the parameters of the shelf configuration on the system flow rates. The analysis is based on the assumption of the flow resistance for other sections of the system remain unchanged. The guideline with the dimensionless form is developed for general application.

The following example will show the calculation procedure about how to utilize the above two figures to obtain the new system flow rate.

Example 6.2

Given: Baseline configuration gives Q0, R1278 = 0.5 Rt and R45 = 0.3 Rt
Find: the new total system flow rate when the inlet/exit height is increased to two times of the baseline configuration and the depth is also increased to 1.5 times of the baseline configuration.

Solution

Step 1 to compute Q1′:
from Figure 6.8, we have Q1′ = Q1/Q0 = 1.27
Step 2 to calculate Q2′:
From Figure 6.9, we have Q2′ = Q2/Q1 = 1.095

The total dimensionless system flow rate Q2 is 1.39 (= 1.27×1.095) Q0 which is 39% higher than the flow rate at the base line configuration.

Example 6.3

A 19 in. telecommunication shelf with 12U height is under consideration. The height of the inlet and exit is 1U. The system includes 14 slots and 12 axial fans. The individual fan power consumption is 24.96 W. The results from the CFD analysis on this baseline configuration indicate that (R18)bl = 0.5447(Rt)bl and (R45)bl = 0.3(Rt)bl. Examine the effects on the system flow rate if the inlet and exit height is increased to 1.5U, which results in the total height of the shelf of 13U (Figure 6.10).

For *H*/Hi = 1.5 and R18 = 0.5447 Rt

From Figure 6.3, we have $Q_{new}/Q_{baseline} = 1.2$.

The result indicates that the flow rate is increased by 1.2 times over the flow rate at the baseline configuration when the height of the inlet and exit is increased from 1 to 1.5U.

The potential opportunities due to the increased the system flow rates are as follows:

1. Reducing the fan speeds to save the energy consumption and achieve the system energy efficiency by keeping the system flow rate at the baseline configuration.
2. Utilizing this increased the flow rate to reduce component temperatures, which leads to higher system reliability. In addition, the results also open the opportunity of using lower component temperature limits of the commercial grade components, which are less expensive.

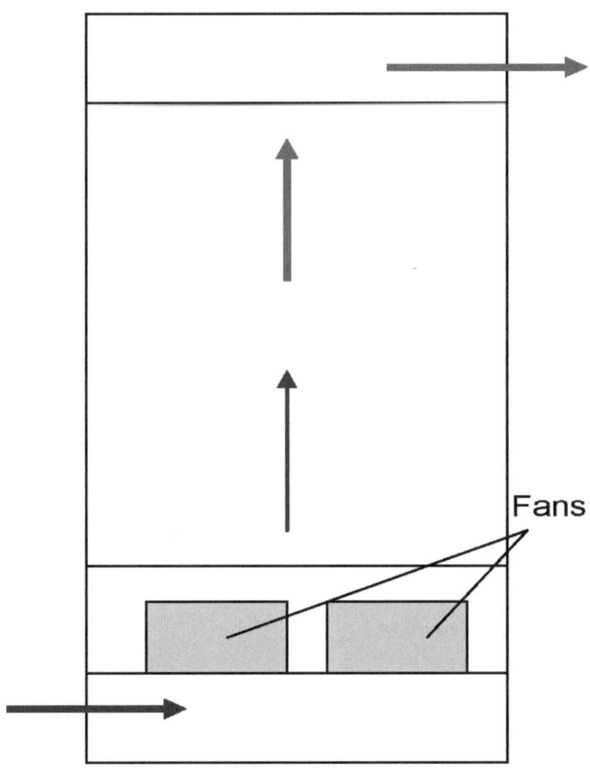

FIGURE 6.10 A telecommunication shelf under consideration.

To examine how much energy can be saved, one must understand the operation of the fan laws given below

$$Q = \Phi N D^3 \tag{6.9a}$$

$$P = \lambda \rho N^3 D^5 \tag{6.9b}$$

$$p = \psi \rho N^2 D^2 \tag{6.9c}$$

The variables involved in the above equations are fan size D, rotational speed N, gas density ρ, volumetric flow rate Q, pressure p, power P, and the fan efficiency η. Variables Φ, ψ, and λ are the constants for geometrically and dynamically similar operation and are also referred to as the flow coefficient, pressure coefficient, and power coefficient, respectively.

To maintain the same $Q_{baseline}$ at H = 1.5Hi, the new fan speed and power consumption can be computed according to the above fan laws as follows:

$N_{new}/N_{baseline} = (1/1.2) = 0.83$
$P_{new}/P_{baseline} = (0.83)^3 = 57.9\%$

Total energy saving = 12 × 24.96 × (1 – 0.579) = 126.1 W.
Total annual energy saving = 365 × 24 × 126.1/1000 = 1104.6 KW Hr.

For D/Di = 1.5 and R45 = 0.3 Rt:

The new configuration under consideration now is that the depth is increased by 50% while keeping H/Hi = 1.5.

From Figure 6.4, we have $Q'_{new}/Q_{new} = 1.1$

$Q'_{new}/Q_{baseline} = (Q'_{new}/Q_{new}) \times (Q_{new}/Q_{baseline}) = 1.1 \times 1.2 = 1.32$
$N'_{new}/N_{baseline} = (1/1.32) = 0.758$
$P'_{new}/P_{baseline} = (0.758)^3 = 0.435$

Total system energy saving = 12 × 24.96 × (1 – 0.435) = 169.2 W
Total annual energy saving = 365 × 24 × 169.2/1000 = 1482.2 kW Hr

The above two examples clearly demonstrate the value of the family curves presented in the above figures which allows obtaining quick solutions.

6.1.4 Accuracy of Results

The above family of curves generated [1] is based on the relationship between the pressure drop and the flow resistance as described by Equation (6.1). The simplified analysis is also based on the assumptions listed in Section 6.1.2. The goal of the present analysis is to provide quick solutions to examine the effects of shelf height and/or the shelf depth on the system flow rates so that the system can be optimized to achieve the energy efficiency.

The CFD analysis has been performed on several cases to evaluate the accuracy of the present analysis [1].

Case 1: R1278 = 0.596 Rt

The height of the inlet and exit of the system under consideration as shown in Figure 6.11 is 1U for the baseline configuration.

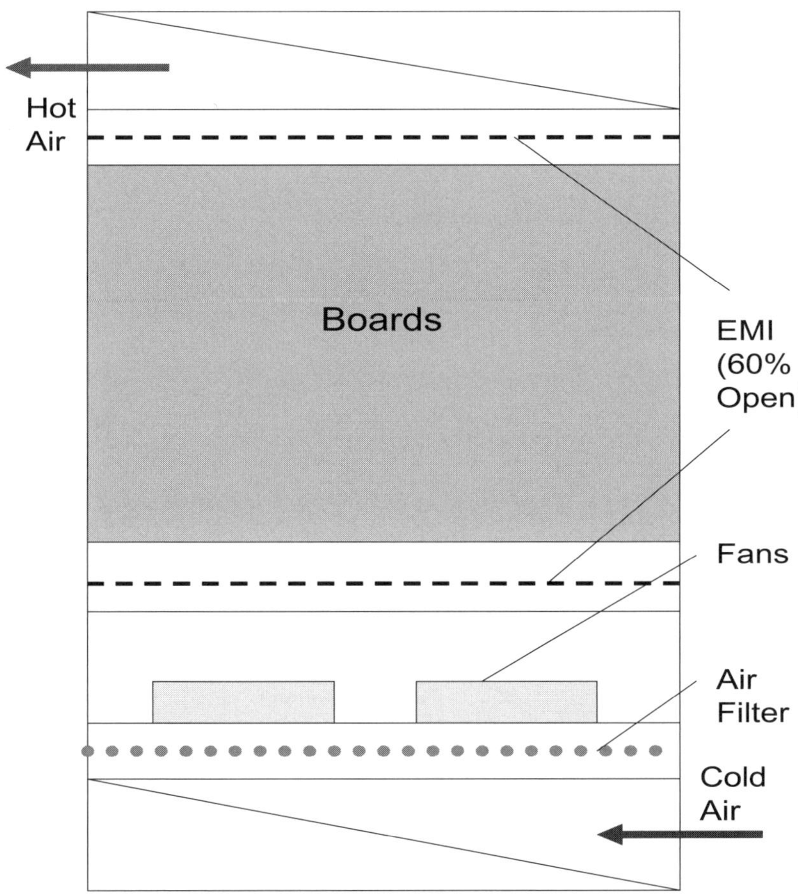

FIGURE 6.11 System under consideration.

The detailed results from the CFD analysis are available for the baseline configuration. Among them, the flow resistance for the inlet/exit section (R1278) is 0.596 Rt. The additional CFD analyses were also performed for the cases with 2U and 3U inlet/exit heights. The detailed CFD analysis results were given in Tables 6.1 and 6.2. The former lists the pressure drop over individual sections of the shelf, and the latter presents the system flow rate and pressure drop.

Table 6.1 CFD—Pressure Drop over Various Sections of Shelf

	Configuration—1U inlet/exit (in of water)	Configuration—2U inlet/exit (in of water)	Configuration—3U inlet/exit (in of water)
Inlet/inlet section +exit/exit section	0.3888	0.2164	0.142
System ΔP	0.6524	0.5808	0.5476
ΔP%	100%	89.0%	83.9%

R1278 for baseline configuration (Configuration – 1U)0.3888/0.6524 = 0.596.

Table 6.2 CFD—System Flow Rate and Pressure Drop

	System Pressure ΔP (in of water)	System Air flow rate (CFM)	CFM ratio over 1U baseline
Configuration—1U Inlet/exit (baseline)	0.6524	287.2	1.0
Configuration—2U Inlet/exit	0.5808	387.5	1.349
Configuration—3U Inlet/exit	0.5476	429.5	1.495

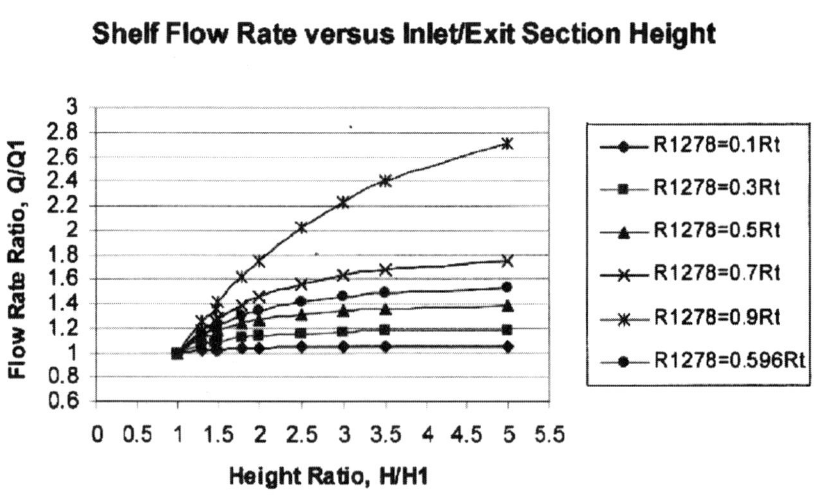

Figure 6.12 Inlet/exit section height versus system flow rate.

Table 6.3 Comparison between CFD and Graphical Results

Height ratio	Graphical flow rate ratio	CFD flow rate ratio	% Difference
1 (1U)	1.0	1.0	0.00
2 (2U)	1.345	1.349	−0.30
3 (3U)	1.458	1.495	−2.47

As can be seen from the above tables, the system flow resistance or system pressure decreases and the system flow rate increases, respectively, when the height of the inlet and exit increases.

Following the calculation procedure described in Section 6.1.1, an additional curve for R1278 = 0.596 Rt is added to Figure 6.8, and the new charts are presented in Figure 6.12. The reason of creating the new curve for R1278 = 0.596 Rt is to get an accurate result from the charts. For practical applications, the predicted results can be obtained by the interpolation between curves for R1278 = 0.5 Rt and R1278 = 0.7 Rt from Figure 6.8.

The graphical flow rate ratios over the height ratios from the above figure are given in Table 6.3. Table 6.3 also includes the CFD results which are obtained from Table 6.2. The graphical representation of Table 6.3 is presented in Figure 6.13.

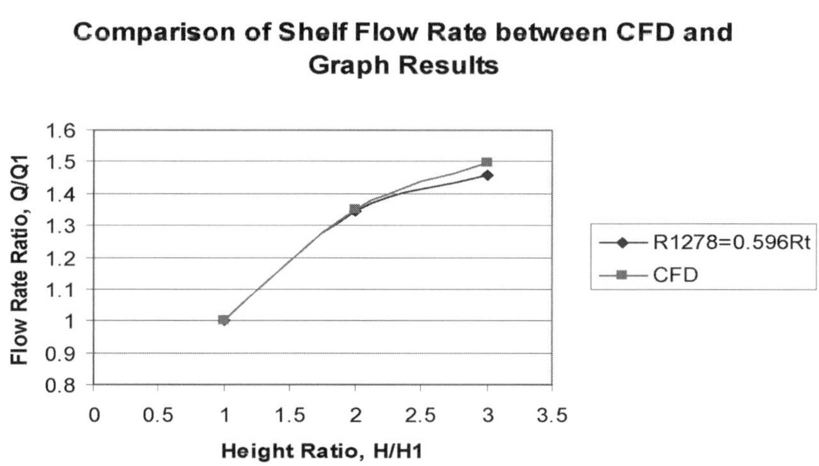

Figure 6.13 Comparison between graphical and CFD results for R1278 = 0.596 Rt.

Case 2: R1278 = 0.237 Rt

Another case with R1278 = 0.237 Rt was analyzed. Similar to Case 1, an additional curve for R1278 = 0.237 Rt is added to Figure 6.12 as shown in Figure 6.14 in order to get an accurate result.

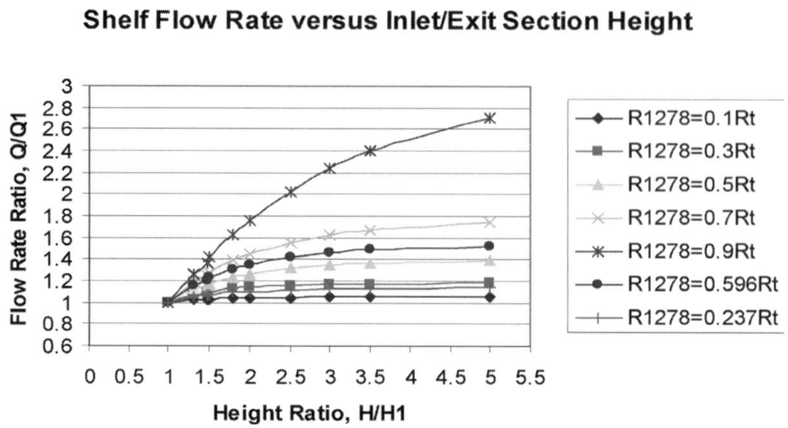

FIGURE 6.14 Inlet/exit section height versus system flow rate.

The comparison between the CFD and graphical results are presented in Table 6.4 and Figure 6.15.

Table 6.4 Comparison between CFD and Graphical Results

Height ratio	CFD results	Graphical flow rate ratio	CFD flow rate ratio	% Difference
1 (1.4U)	410 CFM	1.0	1.0	0.00
1.47 (2U)	434.5 CFM	1.067	1.060	0.66
1.79 (2.5U)	440.9 CFM	1.093	1.075	1.67

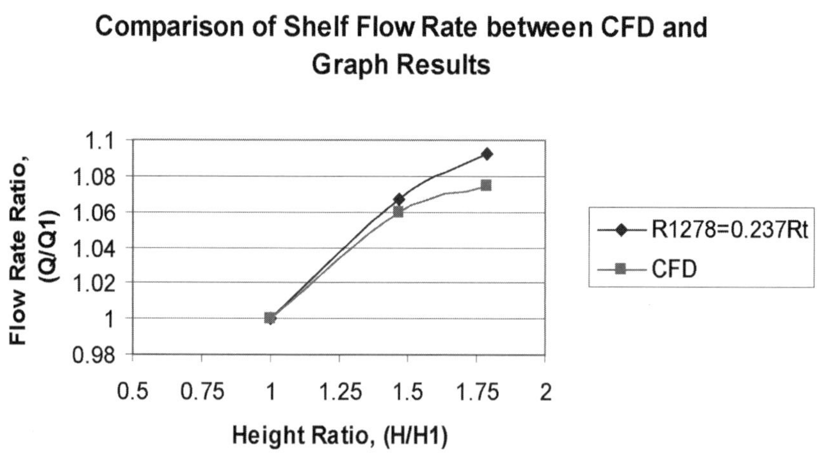

FIGURE 6.15 Comparison between graphical and CFD results for R1278 = 0.237 Rt.

Case 3: R5 = 0.44 Rt

This case is to study the effect of the shelf depth on the system flow rate. As before, an additional curve for R5 = 0.44 Rt is added to the charts in Figure 6.7, and the charts are shown in Figure 6.16. The comparison between the CFD and graphical results is presented in Table 6.5 and Figure 6.17. Again, it should be noted that the graphical results can be obtained by interpolation between curves for R5 = 0.3 Rt and R5 = 0.5 Rt.

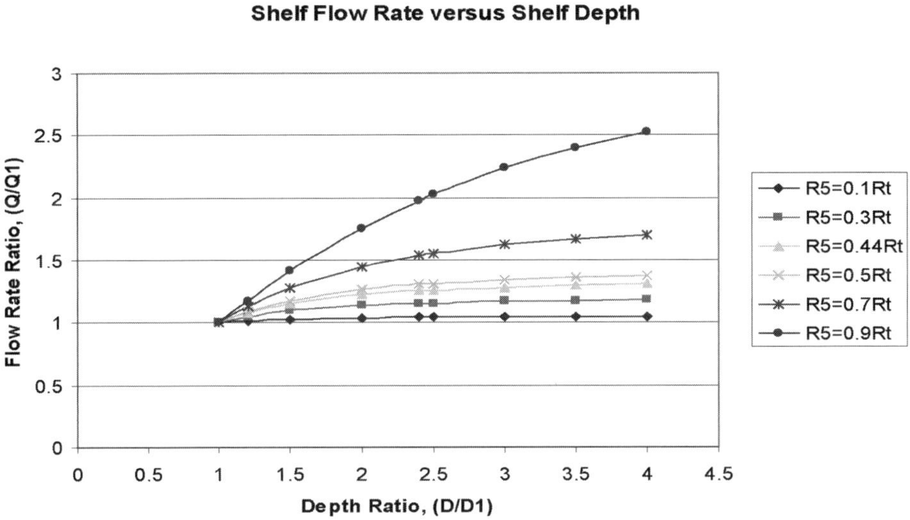

FIGURE 6.16 Shelf flow ratios versus shelf depth ratios.

It seems from Figure 6.17 that the difference between the CFD and graphical results increase as the shelf depth ratio increasing. This is possibly due to the assumptions made in the analysis. One of the assumptions states that the flow resistance of other sections or parts remains unchanged. This assumption is no longer valid if the depth ratio is very large. In summary, it is seen from the above three case studies that the graphical and the CFD results are in an excellent agreement.

Table 6.5 Comparison between CFD and Graphical Results

Depth ratio	CFD results	Graphical flow rate ratio	CFD flow rate ratio	% Difference
1 (10 in.)	547.1 CFM	1.0	1.0	0.00
1.2 (12 in.)	598.1 CFM	1.075	1.093	−1.65
1.5 (15 in.)	623.8 CFM	1.150	1.140	0.88
2.0 (20 in.)	640.0 CFM	1.222	1.170	4.44
2.4 (24 in.)	644.4 CFM	1.254	1.178	6.45

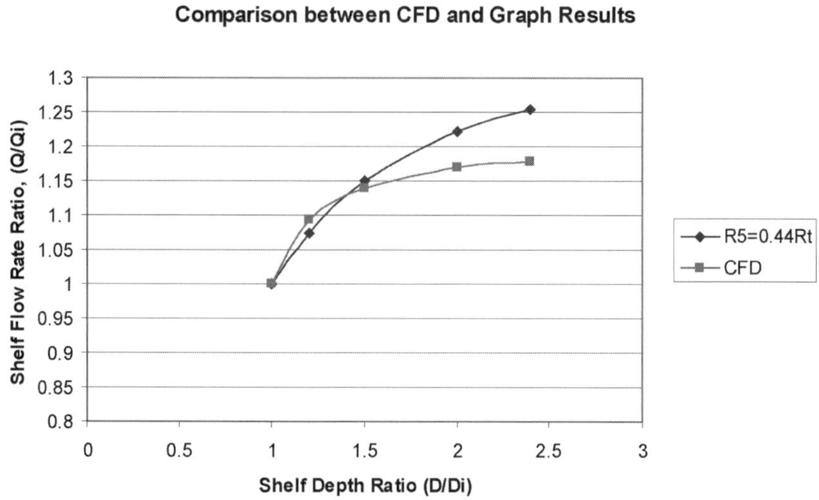

Figure 6.17 Comparison between graphical and CFD results for R5 = 0.44 Rt.

6.2 VARIOUS SUBJECTS RELATED TO SYSTEM THERMAL DESIGN

There is no "one size fits all" design approach. This section further discusses the effects of various parameters on thermal design of telecom systems.

6.2.1 Effect of Series Fan Trays

It is a well-known fact that the fans in series will provide high available head (pressure), while fans in parallel will give a high flow rate. The typical fan performance curves for the fans in series and in parallel are illustrated in Figure 6.18 [2]. It should also be noted that the flow rate is always increased for the fans in series. However, the rate of increase decreases as the number of the fans increases.

It is of interest to also examine the effect of the fan trays in series. It should be noted that the fans are arranged in parallel in the fan tray. Two cases, as shown in Figure 6.19, are under consideration. The number of the fan trays is increased from one to three in both cases. However, each shelf has one fan tray in Case B. The flow rates from the CFD analysis are normalized with respect to one fan tray case. The comparison between Cases A and B are presented in Figure 6.20. The rate of increase for Case A is generally in agreement with the expectation. On the other hand, the flow rate in Case B is only increased slightly as the number of fans increases. The main reason is that the length of the flow path is also increased with the number of the fan trays in Case B.

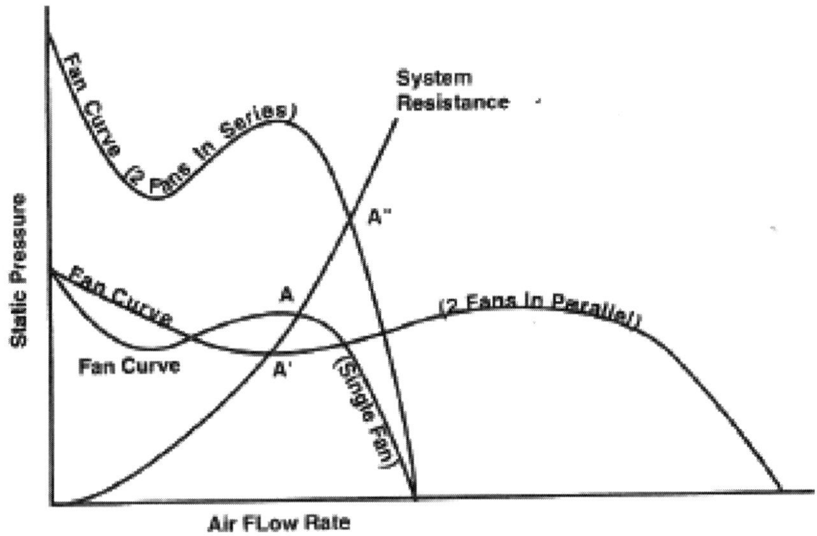

FIGURE 6.18 Performance curves of fans in series and parallel.

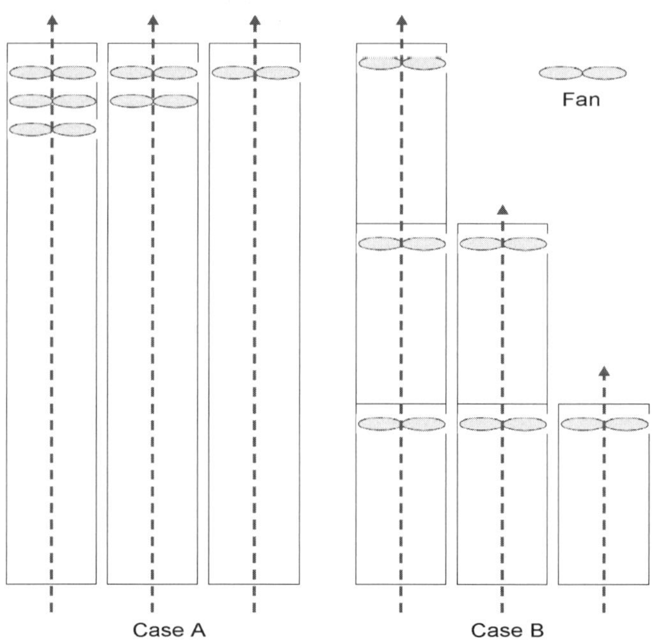

FIGURE 6.19 Fan trays in series.

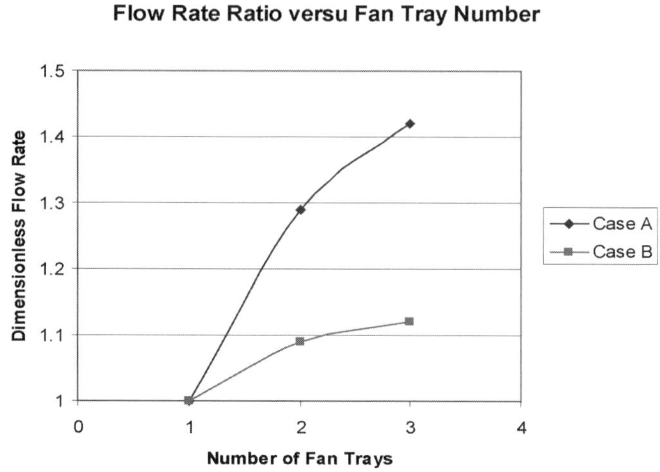

FIGURE 6.20 Comparison of flow rate between Cases A and B.

6.2.2 Factors Affecting Component Temperature

The most critical factors affect the temperature of the components on the board are the heat sink configuration and the airflow rate. However, both factors are too complicated to give some sort of general design guidelines. Therefore, this section examines other factors which are easily identified and may affect the component temperature, while the heat sink configuration and the airflow rate remain unchanged.

Thermal Interface Material (TIM)

The thermal interface material (TIM) is often inserted between the heat sink and the component in order to enhance the heat transfer from the component to the heat sink as illustrated in Figure 6.21. It is a well-known fact that the thermal resistance always exists at the interface when two materials are brought together.

The total thermal resistance across TIM can be expressed as follows [2]:

$$Rt = Rth + Ri1 + Ri2 \tag{6.10}$$

where

Rt = total thermal resistance
Rth = thermal resistance due to TIM thickness (=L/KA)
Ri1 = interface thermal resistance between heat sink and TIM
Ri2 = interface thermal resistance between component and TIM

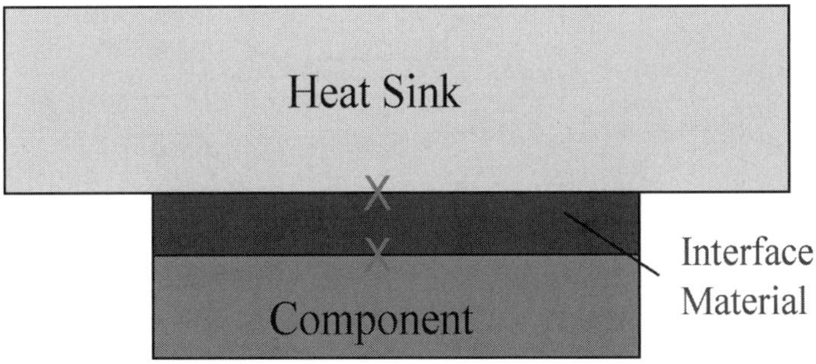

FIGURE 6.21 Sketch of heat sink and component.

Rth is a function of material thermal conductivity, thickness, and void fraction as well as its distribution, of TIM. Ri1 and Ri2, which can only be determined experimentally, depend on bond process and surface characteristics and material properties of Solids 1 and 2. It is difficult to compute Ri1 and Ri2 separately. Therefore, Ri1 and Ri2 can be combined into a single value, and Equation (6.10) can be rewritten as follows:

$$\mathrm{Rt} = \mathrm{Rth}(1 + C/\mathrm{Rth}) \tag{6.11}$$

where *C* represents the combined interface thermal resistance at two interfaces.

i. The *C* value becomes significant when Rth is small; this case implies that thermal conductivity of TIM is very large or the thickness of TIM is very small.
ii. The *C* value is negligibly small when Rth is large; this condition indicates that thermal conductivity of TIM is very small or the thickness of TIM is very large.

Equation (6.10) has been validated by the following experimental data presented in Figures 6.22 [2] and 6.23 [3].

The experimental data indicated that the combined interface resistance *C* could vary from 10% to 80% of the theoretically calculated value (Rth) as epoxy conductivity increased from 0.27 to 1.93 W/m °C. The following correlation was developed by Yeh (2) to predict the total thermal resistance over an epoxy with thickness ranging from 0.018 to 0.022mm:

$$\mathrm{Rt} = 0.26(\mathrm{Rth})^2 + 0.62\,\mathrm{Rth} + 0.15 \tag{6.12}$$

where the unit of the thermal resistance is cm^2 °C/W.

Figure 6.23 shows the thermal resistance as a function of the applied pressure. At low pressure <15 psi, the total interface thermal resistance can be more than 50% of the thermal resistance of the material itself. The delicate leads and

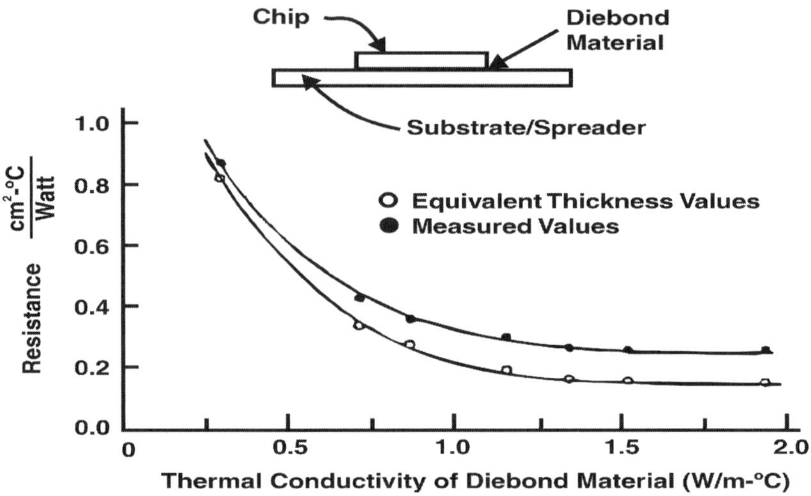

FIGURE 6.22 Measured and calculated thermal resistance of an epoxy joint.

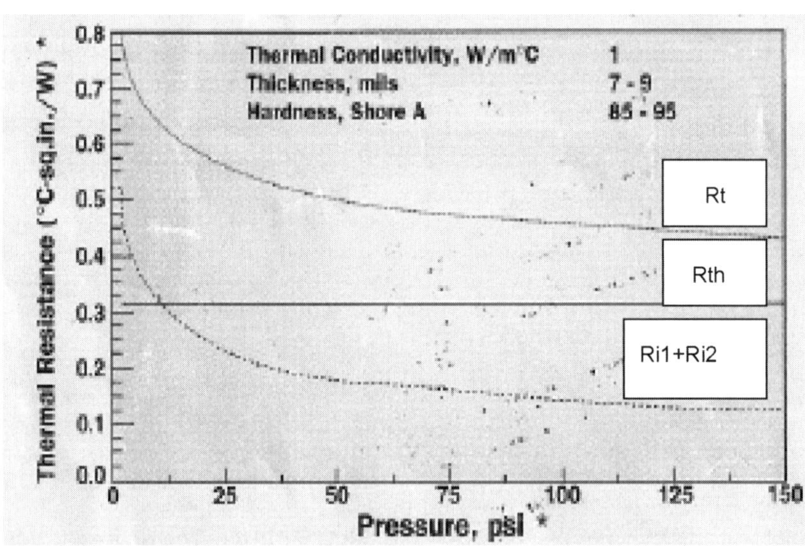

FIGURE 6.23 Thermal pad joint thermal resistance versus pressure.

solder joints can be broken from an excessive pressure. The maximum pressure for most heat sink clip is about 25 psi. However, often 15 psi is tolerated for most components.

There are many types of TIMs, including, but not limiting to, thermal grease, thermal paste, thermal gel, thermal epoxy and thermal pad. The characteristics of thermal grease, thermal gel, or thermal paste are alike and can be lumped into a single group and "thermal grease" will be used as an example for further discussion:

i. Thermal Grease

–Only a small amount of thermal grease should be used, to just fill the void spaces (cavity of rough surfaces) and still allow the high points of the metal surfaces to touch each other.

–Too much grease may create voids within the heat compound and increase the thermal resistance across the thickness of the grease.

–Care must be taken to insure no contamination or any foreign subjects in grease.

–Thermal grease has a tendency to migrate in high temperature and to evaporate in low pressure or vacuum environments. Once vaporized, they may redeposit on adjoining surfaces or disappear altogether.

–It is generally very difficult to apply thermal grease uniformly over a large area (> 3 × 3 in.).

ii. Thermal Epoxy

–Curing temperature and/or pressure may be important to epoxy bonding.

–Voids always exist within epoxy (% of voids depends on application processes).

–Large voids and also void distribution may have a significant effect on thermal resistance over epoxy.

iii. Thermal Pad

–Thermal pad is probably the only TIM that it can be used to address the component height tolerance issue because of its compressibility.

–Thickness required for a thermal pad is generally larger than those of thermal grease or epoxy.

–A large factor in reducing pad thermal resistance is the application of pressure.

–Increasing pressure not only reduces the thermal contact resistance but also can reduce the material thermal resistance by reducing the material thickness.

–Increasing pad thermal conductivity generally will decrease material compressibility.

Figure 6.24 [3] also shows the pressure limits for PQFP CPU and for most heat sink clips. The delicate leads and solder joint can be broken with excessive pressure.

It is concluded that the thermal resistance at two interfaces of a joint cannot be ignored in determining the total joint thermal resistance which is always greater than Rth based on the material's thickness alone.

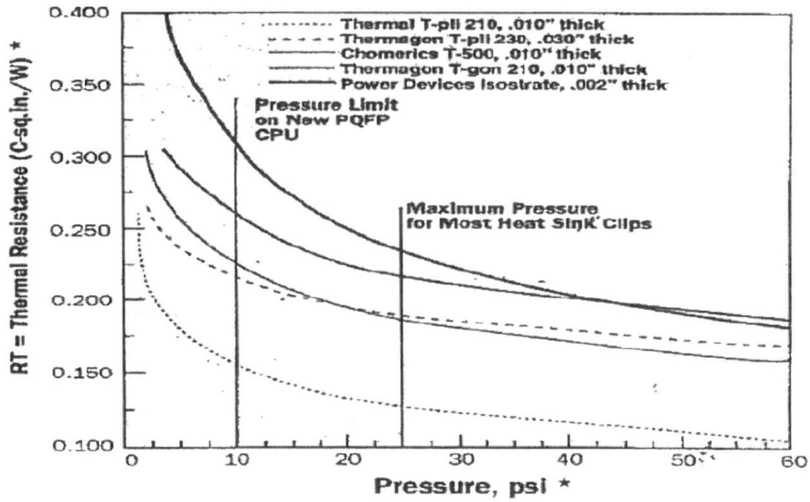

FIGURE 6.24 **Thermal resistance over various pad joints.**

Thermal Conductivity of Printed Circuit Board

The board provides the structure support and electrical connections to all components. However, it is seldom known that the board is the largest heat sink available to the components. Increasing the board thermal conductivity increases the heat conduction spreading along the board and reduces the component temperatures. Therefore, the thermal conductivity of the board has a great effect on the component temperature.

The thermal conductivity of the board depends on the construction and materials of the board. To improve the board thermal conductivity, one may consider to employ polyimide in replacing of FR-4 in the board construction because the conductivity of polyimide (k = 0.52 W/m °C) is about 50% higher than that of FR-4 (k = 0.35 W/m °C). The practical approach in the thermal analysis is to determine the so-called equivalent thermal conductivity for the *XY* plane and the *Z* plane which are function of the copper contents.

Baffles in Inlet/Exit Section

The baffle (often is also referred to as deflector) as shown in Figure 6.25 is used to eliminate or minimize the flow recirculation zones in order to improve the system flow. The effect of the baffle on the system flow in the forced air-cooled system with fans is relatively limited. The main affect of the baffle is slight changes of the flow distribution over the board which may have some effects on the component temperatures. The baffles increase the airflow distribution to the rear section of the board and therefore slightly decrease the temperature of those components closer to the back plane. To be effective, the baffle must be added to the inlet and exit section of the shelf simultaneously.

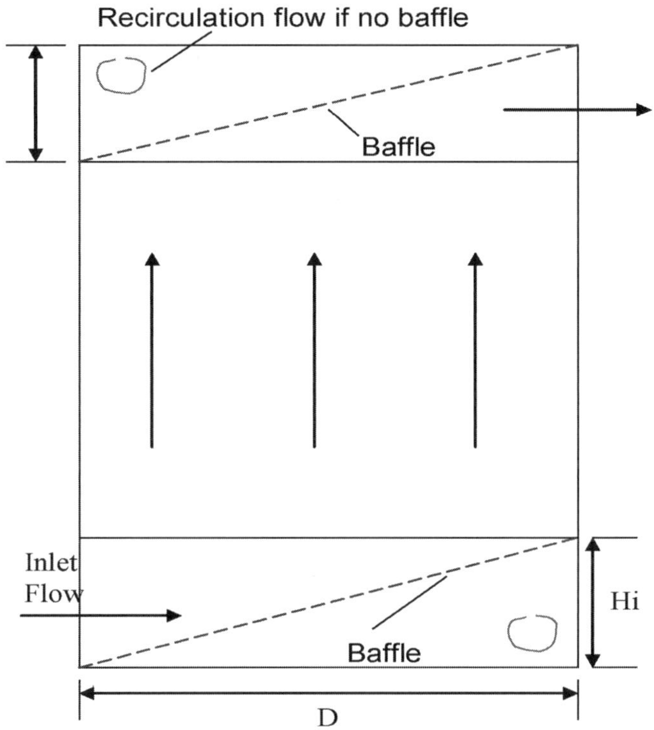

FIGURE 6.25 Sketch of shelf with baffles.

However, the effect of the baffle is much profound in the natural convection environment (no fan condition). In other words, the shelf flow rate will be increased in the case without the baffles. This is due to the fact that the baffle increases the system pressure drop. The available pressure head is pretty much fixed in the natural convection conditions once the system height and the heat load are defined.

Fillers in Card Cage

The understanding of the effect of the empty slots in the card cage on the thermal analysis and the tests is extremely important. The detailed full scale thermal model of the rack is sometimes too large for the practical applications. Similarly, the fillers, which are often referred to as the dummy boards, are frequently required during the system thermal tests to insure the proper flow distribution to all slots (or boards).

The system under consideration is a low profile shelf as given in Figure 6.26. The CFD analysis is performed for both cases; one with fully filled card cage (Case A) and another with three-board card cage (Case B) as shown in Figure 6.27. The

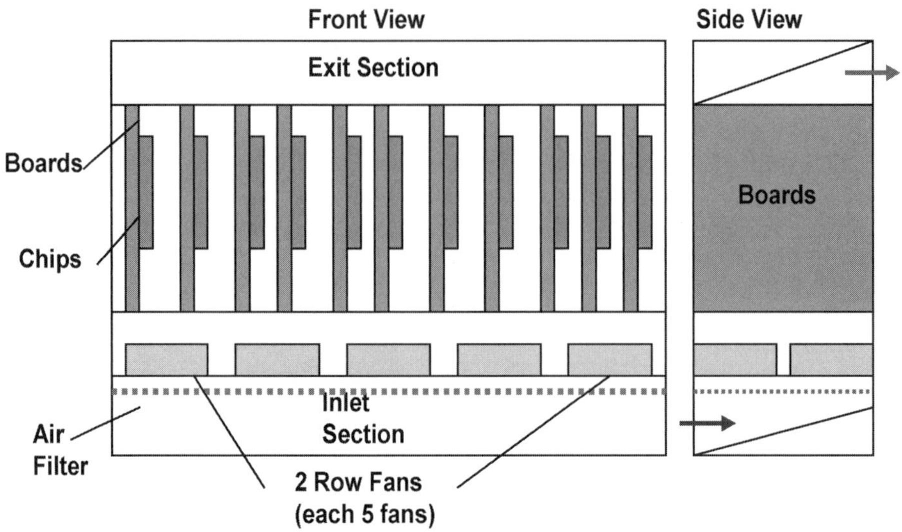

FIGURE 6.26 System under consideration.

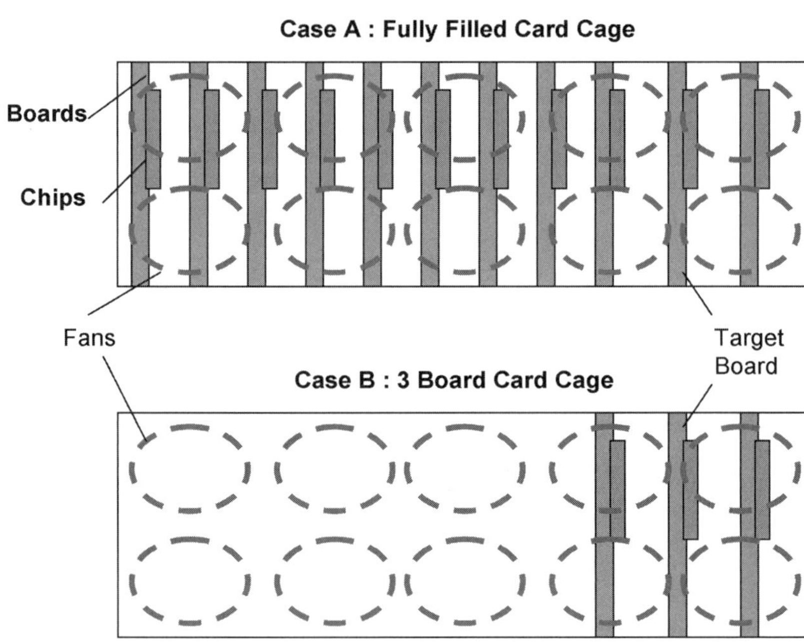

FIGURE 6.27 Top view of fully filled card cage and three-board card cage.

target board is sandwiched by two neighbor boards. This arrangement is needed to insure the accurate results for the target board which will be affected by the neighbor boards. There is no noticeable difference in the predicted results for the target board between Cases A and B (with empty slots in the shelf) in the normal operation (all fans on) and one fan failed condition.

The results indicate that the target board is very insensitive to the operation conditions in far away locations. In other words, the results of the target boards are relatively independent of the performance of the other boards and fans located at the distance from the boards under consideration. The main reason is that the board thermal performance is generally governed by the fans directly located under or above the boards under consideration. In this particular case, the fans are located so closed to the card cage that the airflow distribution to the boards is directly linked to the fans right below the board under consideration. As required by GR-63, the components should also meet the respective temperature limits under the one fan failure. The effect of one fan failure is generally a local phenomenon and only the board direct above or below the failed fan will be affected.

The above finding is limited to the forced air cooling with the fans in the system. The attention is now turned to the equipment without the fan. The components on the board in such a system are subjected to the thermal radiation and natural convection. The passive cooling is simple, cost effective and reliable. The system under consideration as shown in Figure 6.28 is a 4U shelf, which includes fifteen 2U boards. The shelf without the fan is cooled by natural convection. Thermal radiation effect must always be included in the thermal analysis in a passive cooling condition.

The purpose of the analysis is to examine the effects of the empty slots or the fillers on the temperature of the components on the target board. The CFD analysis is first performed on the fully heated card cage and the results are given in Figure 6.29. The maximum component temperature for the target board is 122°C and for two neighbor boards are 112°C and 114°C for Cards A and B, respectively. These results are considered to be baseline values.

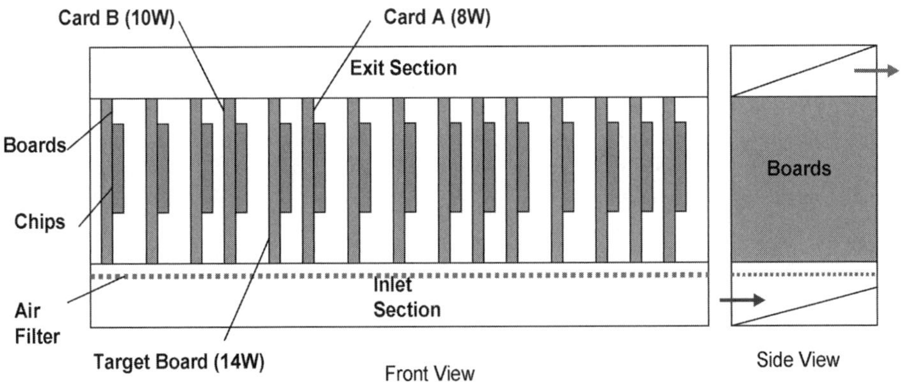

FIGURE 6.28 A passive cooled 4U shelf.

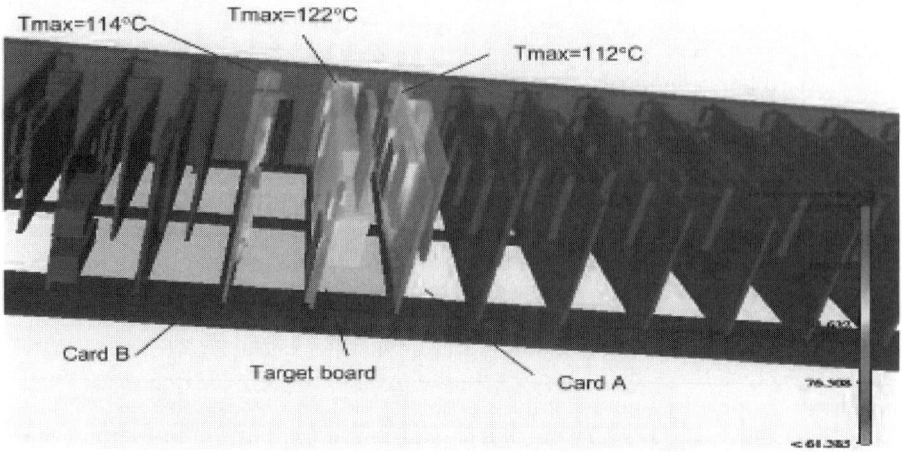

FIGURE 6.29 Results for fully heated boards in shelf.

The next step is to turn off the power to all the boards with exception of the target board and Cards A and B. The results for the three-heated board in the fully filled card cage are given in Figure 6.30. Due to radiation, heat is transferred from the heated boards to the zero power boards which results in significant reduction in the temperature of the components on the three powered boards.

The analysis is further extended to the case without the zero power fillers in the card cage. In other words, all zero power cards in the above case are removed, and the card cage only has three boards plus a lot of empty slots. The component temperatures on the target board and Cards A and B are presented in Figure 6.31. As can be seen from the figure, the component temperatures are reduced significantly due to the empty slots. The reasons for the lower component temperatures are because one side (opposite to the target board) of Cards A and B is basically exposed to the ambient air induced by the natural convection and also radiation to unheated structure in the surrounding.

Examining all results given in this section, the effects of fillers can be summarized as follows:

1. Forced Air Cooled Systems
   No noticeable difference in the component temperatures between the cases of the fully loaded and partially filled card cages under the force air cooling with the fans.

   The main reason is that no mixing of airflow from fan exits due to the limited space existing between the card cage and the fan tray. Furthermore, the thermal performance of the individual boards is generally governed by the performance of the fans directly located below or above.
2. Passive Cooled Systems
   No significant difference in the component temperatures between the cases with the zero power fillers and the empty slots under the natural convection conditions. However, both cases significantly reduce the component

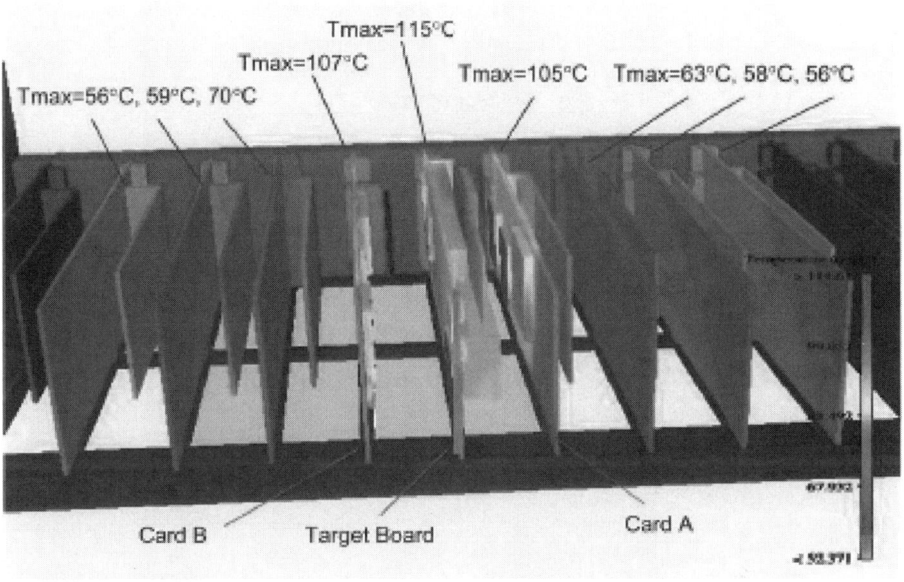

FIGURE 6.30 Results for three-heated-board card cage with zero power on fillers.

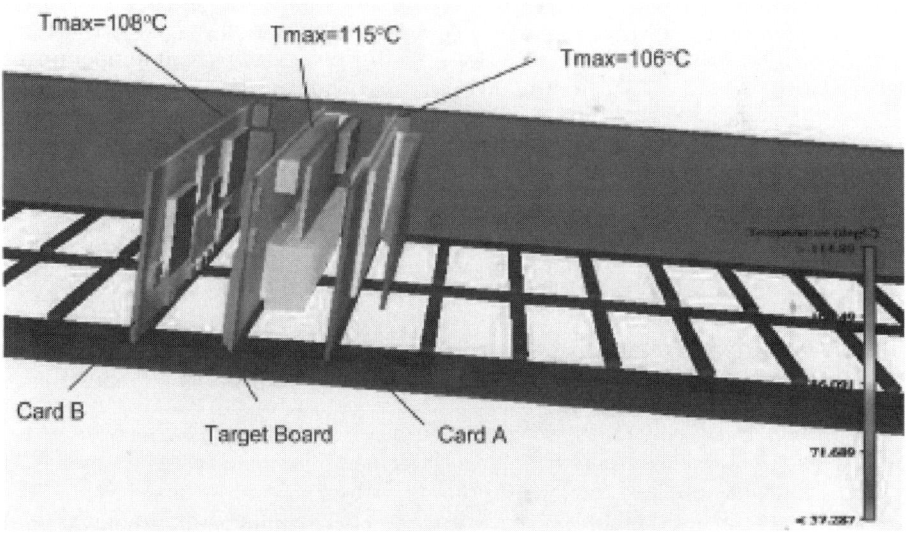

FIGURE 6.31 Results for three-heated-board card cage without fillers.

temperatures of the heated boards as compared with the fully loaded card cage. Therefore, the thermal simulations and/or tests for the passive cooling should be conducted under the actual operation conditions with the full power on all boards in the card cage.

6.3 THERMAL ENHANCEMENT

To meet the component temperature limits, various methods can be considered to reduce the temperature of the components on the board. The typical steps which can be taken to improve the thermal performance of the telecommunication racks are described in the following section.

1. Board Level:
   - (i) Optimizing heat sinks and board layout
   - (ii) Enhancing board flow management (e.g., adding baffles to guide flow)
2. System Level:
   - (i) Optimizing system flow path for minimizing system ΔP in order to increase flow rates
   - (ii) Utilizing perforating plates or baffles in inlet and/or exit section to regulate the proper flow rates to the boards
   - (iii) Increasing the surface area of the air filter
   - (iv) Upgrading the fans without changing the size of the fan tray if possible
   - (v) Increasing the height of inlet and/or exit section of the system if possible

Among all above factors, increasing the height of inlet/exit sections may be the simplest way to do with most benefits as long as the total height of the rack remains less than 7 ft which is the limited for North American market.

One critical thermal issue for the current high power and dense packed board is the space available on the board. The large oversized heat sinks needed to accommodate high power components further compound the thermal problems. The modularized board concept as presented in Figure 6.32 is developed to address the above problems.

The unique feature of the modularized board is to add a thin metallic cover which provides an extra large surface for heat transfer in air cooled systems or interfaces with the cold plates in liquid cooled equipment. Other advantages of the cover are as follows:

1. Protection for components during handling and/or shipping
2. Prevention from fire spreading between boards to meet GR-63 requirement

As noted previously, the board is the largest heat sink to all components on the board. To utilize this large heat sink efficiently, it is recommended to place as many low profile and low power chips on the bottom side of the board as possible (up to 15% to 20% of total power of the board). This arrangement not only lowers the surface heat flux on the top side of the board because of reduction of the number of the components but also provides additional space needed for mounting

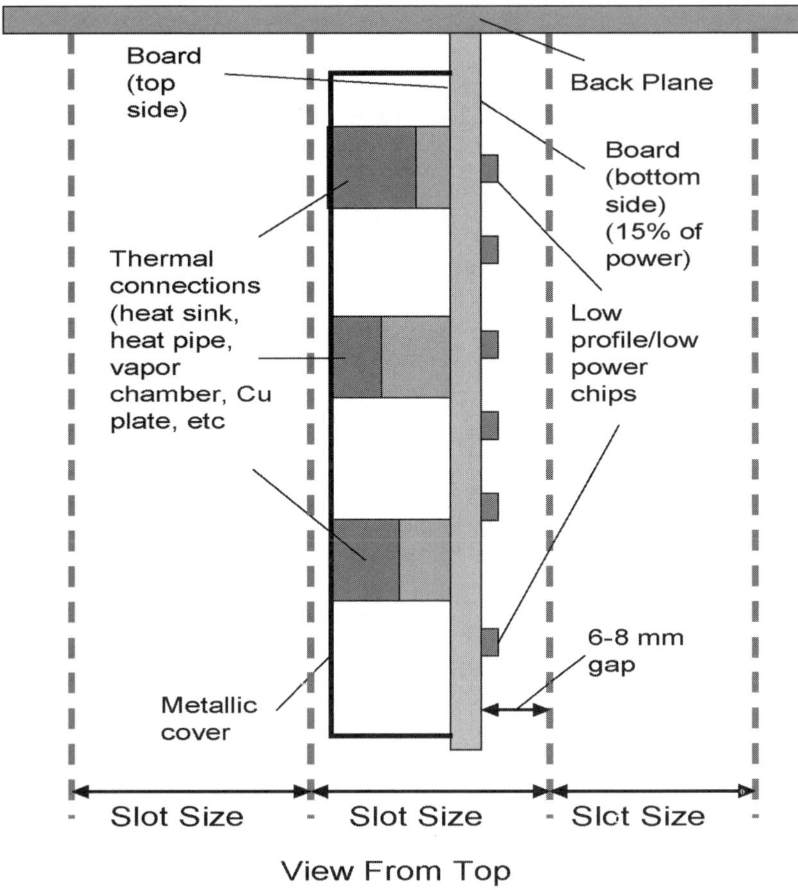

FIGURE 6.32 Modularized board.

the heat sinks. The 6 to 8 mm gap on the bottom side of the board also give the enough space for adding the low profile heat sink, including thin metal plates to the components as needed.

As stated in the above section, the board cover can be utilized as the heat sink to all components on the board if needed. Along this concept, a single large heat sink can be adapted to cover multiple components as shown in Figure 6.33. The current practice for attaching an oversize heat sink to the component is that the heat sink is mechanically mounted to the board which requires punching holes on the board. The holes on the board will greatly affect the board routing. Generally, four holes are needed to mount a heat sink to the component.

For a typical high power board, the board is populated with the components, active and passive components. There is a limited space available for mounting heat sinks. The single large heat sink will reduce the number of holes needed on the board. It will require 12 heat sink holes on the board for three chips as

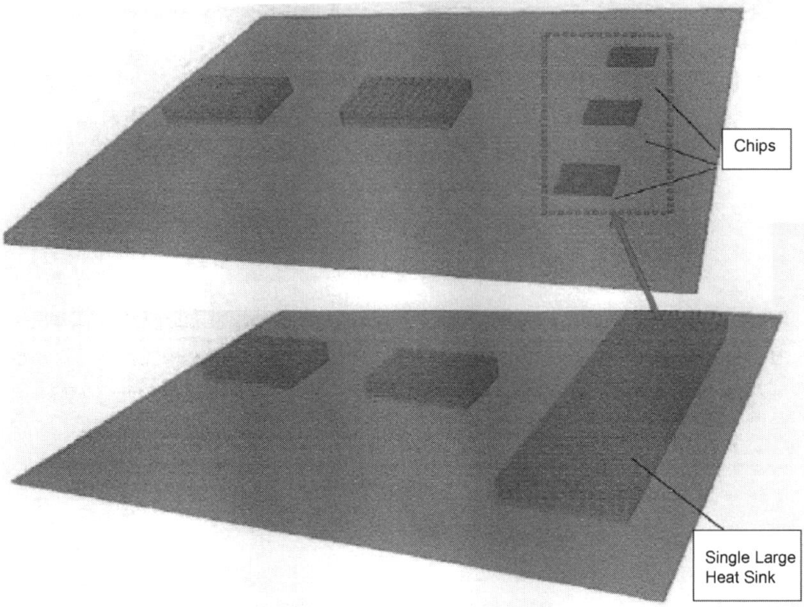

FIGURE 6.33 Large heat sink for multiple chips.

illustrated in Figure 6.33. However, the number of heat sink holes on the board will be reduced to no more than eight if a single large heat sink as shown in the figure is employed. In addition, the single large heat sink will also make chip temperatures more uniform due to heat spreading along the base of the heat sink.

6.4 THERMAL CONSIDERATION IN DATA CENTER OR CENTRAL OFFICE

The inlet and/or exit location of the airflow for a telecommunication system generally does not affect the system thermal performance for stand alone equipment. However, most telecommunication equipment is either deployed in a data center or central office which hosts hundreds of similar equipment. The inlet and/or exit location of the airflow will have some impacts on the system thermal performance due to present of neighbor systems.

Rows of rows of the systems are placed in the data center as shown in Figure 6.34 which represents a typical data center layout. As can be seen, the air inlet is located at the front side of the equipment facing to the cold aisle, and the hot air exits from the rear of the system to hot aisles. CRAC stands for computer room air conditioner. The system with the side inlet and exit is becoming unacceptable in the today's data centers because the hot exit air will significantly effects on the thermal performance of the neighbor racks.

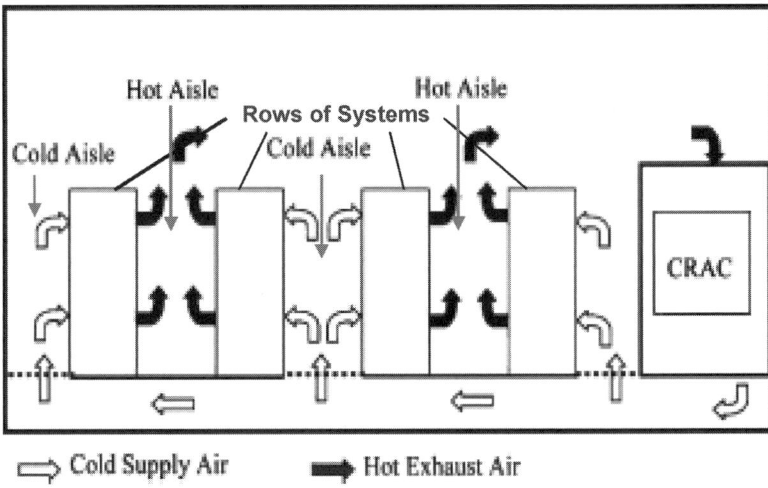

Figure 6.34 Equipment layout in data centers.

The cold air entrance at the top of the system should be avoided because the some of hot exhaust air in the hot aisle will circulate to the front of the rack as illustrated in Figure 6.35. Additional devices such as blanketing panels will be needed to prevent the hot air from circulating back to the front of the rack if the cold air inlet locates at the top of the racks.

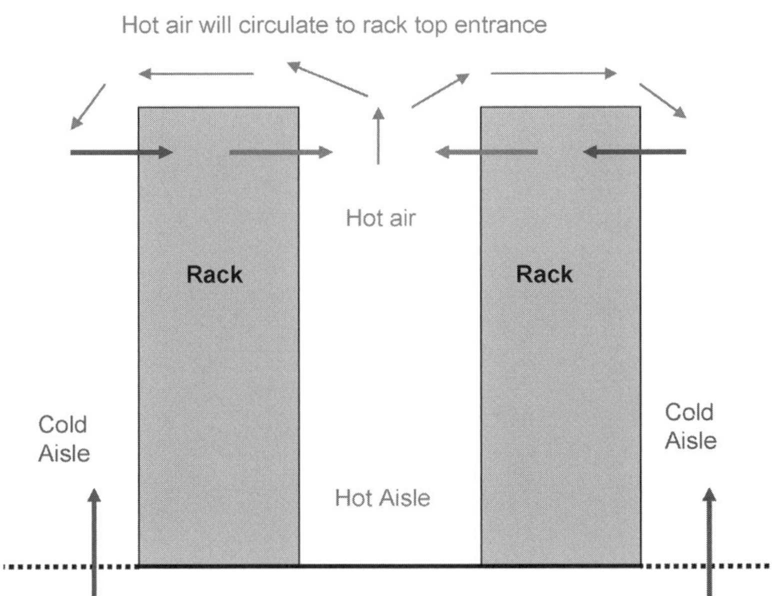

Figure 6.35 Hot exit air circulation.

REFERENCES

1. Yeh, L. T., "Effects of System Height and/or Depth on System Flow Rates for Telecommunication Racks", to be published.
2. Yeh, L. T. and Chu, R. C., Thermal Management of Microelectronic Equipment, ASME Press, 2002.
3. Latham, C. A., "Thermal Resistance of Interface Materials as a Function of Pressure", Electronic Cooling, Volume 2, No. 3, 1996.

Chapter 7

SYSTEM THERMAL ANALYSIS — SMALL BOX

The typical telecommunication boxes for the indoor applications are small with the height less than 4RU (7 in.). The "RU" stands for the rack unit which is 1.75 in. high. However, for simplicity, "U" is commonly used to replace "RU." One of the unique features of these typical boxes is the printed circuit boards being in horizontal positions. As in the rack configurations, the cooling schemes of the system can be either force air cooling with fans or passive cooling under natural convection.

The challenges of the thermal design in the box configurations, especially for low profile box are due to the limitation of the space available and the boards being installed horizontally. The latter is especially critical to the natural convection environments.

7.1 ACTIVE COOLED BOXES

For a forced air cooling, the fans are typically placed on one side of the box. The cold air enters from one side and hot air leaves at the opposite side of the box. The design and operation are still subjected to GR-63 requirements.

As in any actively cooled indoor equipment, the air filter is generally required. Air inlet and exit are through a perforating plate and the percentage of opening is generally less than 65% in order to meet the EMI (electromagnetic interference) requirements. In addition, the percentage of the opening must also meet the structural requirements of the box.

Figure 7.1 shows a typical example of a 19 in. (width) box. The 19 in., which is the most common one used in telecommunication industry, is a standardized frame or enclosure for mounting telecom equipment. The 19 in. space includes the mounting posts. The actual width of the 19 in. equipment is 17.5 in. (444.5 mm). The depth of the box ranges from 220 to 300 mm. The height of the box can be varied between 1 U and 4 U.

The fan selection is extremely important for the low profile box. Ultimately, the fan performance must meet the system thermal requirements. The selection of the fan is also limited by the available space inside the box. The size and the number of printed circuit boards (PCBs) are system dependent. For example, eight small PCBs as given in Figure 7.1 can be replaced by 4 large PCBs or any combination

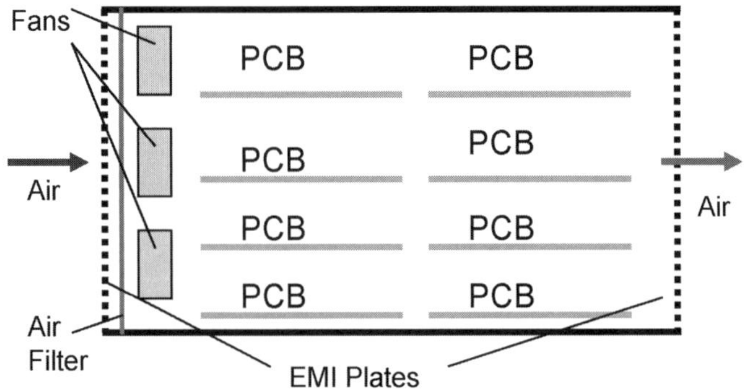

FIGURE 7.1 Sketch of forced air cooling of box.

of small and large PCBs. Another design requirement is that the boards must be able to be removed or plugged into the box freely. The minimum required airflow rate (CFM) is in general about one fifth of the system power in watts. It should be noted that the required airflow rates for a given system strongly depend on the power and types of individual components as well as the board layout (component placement).

The major thermal concern for the small box configuration is that the fans are too close to the boards due to the limited space inside the box. Therefore, the thickness of the fan becomes an issue. The current system employs 28-mm-thick fans. If the system presented in figure is employed to the 4U shelf, the 80-mm fans can be used for such cases.

The location of the fans is actually also depending on the system configuration, especially the available space. Another example of force air cooled box is presented in Figure 7.2. The system under consideration is a 1-U box with the fans installed at the rear of the box. The box contains a large board with various types of components on the board. The fans in the current system are installed in the rear of the box which draws air from the vents at both side walls as shown in Figure 7.2. Sometimes, the ambient air can also be drawn from the perforation at the top of the box to enhance the thermal performance of the box.

In short, the placement of fans is system dependent and should be placed at the least system resistance path. In addition, heat sinks are generally required for most components, especially for those high power components.

Though the boards in the horizontal position are the most common cases in a low profile box, the vertical boards as shown in Figure 7.3 have been adapted. However, the height of the boards is limited to less than 2 U in order to make room for the fan tray, and inlet and exit sections. The total height of the box would probably be increased from 4 U to 5 U, as illustrated in Figure 7.4. In order to increase the board surface area for mounting the components, the subboards or daughter boards are often attached to the main boards. This design can support the system power up to about 350 W.

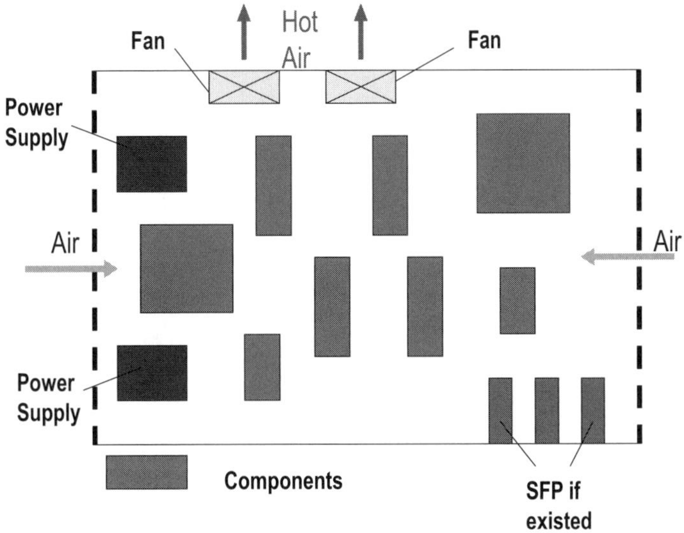

FIGURE 7.2 1 U box with forced air cooling.

7.2 PASSIVE COOLED BOXES

Heat transfer by natural (or free) convection has long been considered as one of the most cost effective and reliable cooling methods. Natural convection with air has many practical engineering applications and is of special interest to the cooling of electronic equipment. The advantages of air cooling by natural convection are simple, reliable, and cost-effective.

It should be noted that the fan tray in Figure 7.3 or Figure 7.4 can be removed, and the system becomes pass cooled equipment. Under the natural convection

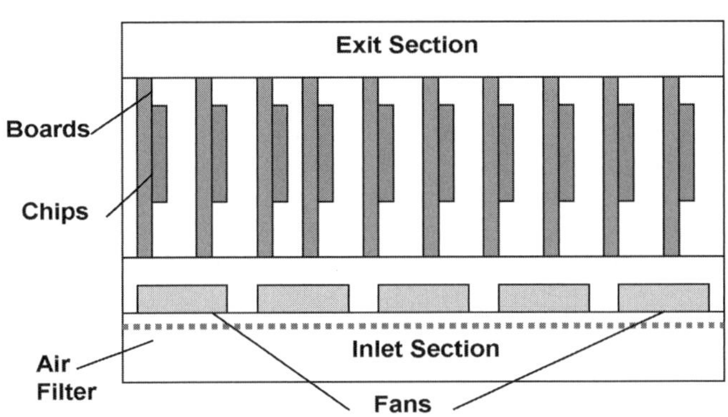

FIGURE 7.3 Vertical boards in low profile shelf.

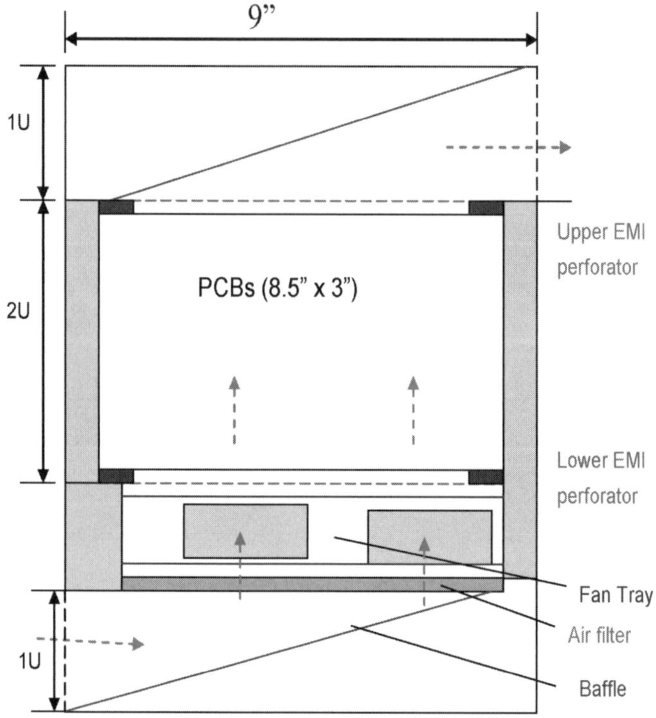

FIGURE 7.4 Cross-section view of vertical boards in low profile shelf.

cooling, the system power as described in the above figures is reduced from 350 W at forced air cooling to about 150W and the system height from 5U to 4U. Furthermore, the inlet and exit sections in the above figures can also be removed for the case with or without fan tray under some limited conditions where sufficient spaces above and below the box under consideration are available to minimize the effects from the surroundings on the system.

The 1-U horizontal box under the natural convection is typically applied to the low power systems in the past. With increasing speeds and functionalities, the system power for a 1 U box has increased significantly in the recent years. Because of the increasing demands on the system performance and reliability, the thermal design is also becoming even more critical and important.

The overall dimensions of a typical 1 U box are 17.5 in. (444.5 mm) (width) × 1.75 in. (44.5 mm) (height) with the depth ranging from 8.67 in. to 11.81 in. (220 to 300 mm). The box is generally mounted to an open telecommunication rack or cabinet, as illustrated in Figure 7.5.

7.2.1 Market Survey

The summary of the survey of the current market for the passively cooled 1 U horizontal box in telecommunication industry is given in Table 7.1. Table 7.1

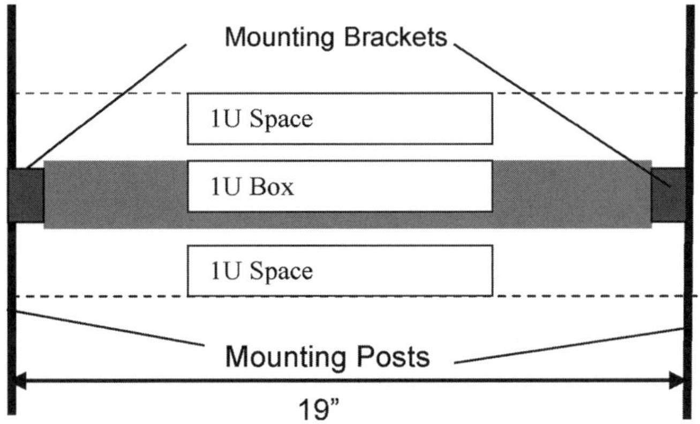

FIGURE 7.5 1 U box in telecommunication rack.

includes the overall dimensions of the individual boxed. As can be seen from Table 7.1, all products in the current market can only support the system power around 50 W. The volumetric power density (W/L) is often used to evaluate the thermal efficiency of the equipment in the natural convection as COP (coefficient of performance) in the forced air cooled systems.

Passively cooled boxes are generally applied to 1 U or 2 U boxes which severely limit the available space inside the box. In addition, the horizontal boards make natural convection even worse. Most of products as mentioned in the above table are in a sealed box environment with a few exceptions having some venting holes on the top surface of the box. The heat generated from all components is first transferred to the enclosed walls and then heat is finally rejected from the walls to the ambient by natural convection and radiation. To enhance the heat transfer, the perforation can be made on the top cover and side walls as NEC Pasolink NEO box. The opening on the top cover is generally limited to less than 60% in order to meet the EMI requirements. The percentage of openings on the sidewalls should be much smaller due to the structural requirements. Another way to enhance heat transfer from components to the cover is often done by applying solid aluminum blocks which are integrated part of the cover directly to components as Fujitsu Flashwave 4020.

Table 7.1 Various Products of 1 U Box in Current Market

Products	Overall dimensions	Power	Power density	Ambient
	(W × D × H) (mm)	(W)	(W/l)	(°C)
Ericsson Mini-Link HC	483 × 280 × 44.45	45.0	7.49	–20–60
NEC Pasolink NEO	482 × 240 × 44.45	47.0	9.14	–5–50
Fujitsu Flashwave 4020	444.5 × 228.6 × 44.45	43.5	9.63	0–50

7.2.2 Case Studies

The system under consideration is given in Figure 7.6. The overall dimensions of the box are 444.5 mm (W) × 246.4 mm (D) × 44.5 mm (H) and the total system power is 53.13 W. The dimensions for the printed circuit board (PCB) are 378 mm (W) × 190 mm (D) 1.6 mm (thickness). The component temperature limit which is also the thermal design limits is generally at 125°C with exception for optical components such as SFP/XFP.

In addition, the following assumptions are made to facilitate the thermal analysis

a. The box is suspended in an open air space
b. The ambient air is at 55°C and at the sea level condition
c. The box is limited to passive cooling with natural convection and radiation
d. The box is made from anodized aluminum with emissivity of 0.81

Item (a) actually implies that effects from other equipment located above and/or below the current system under consideration are small and can be omitted. This may not be always true in the actual installation and this issue of environmental effects should be considered in the system design. Since no active cooling is available, this 1 U horizontal box is subjected to the passive cooling, including the natural convection and radiation. The analysis has been performed on various configurations with aids of Flotherm software [1].

The ultimate goal of the thermal design is to maintain all component temperatures below the respective temperature limits in order to achieve the desired system reliability and performance. The heat is transferred to the ambient through the box surfaces by the natural convection and radiation. The radiation must be included in the case of the passive cooling condition.

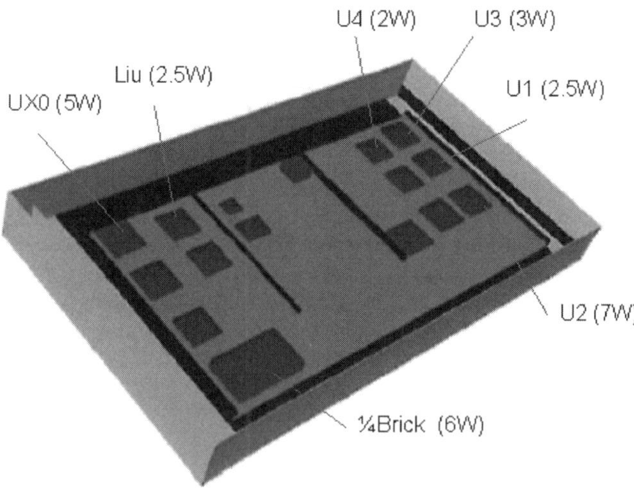

FIGURE 7.6 Box internal layout.

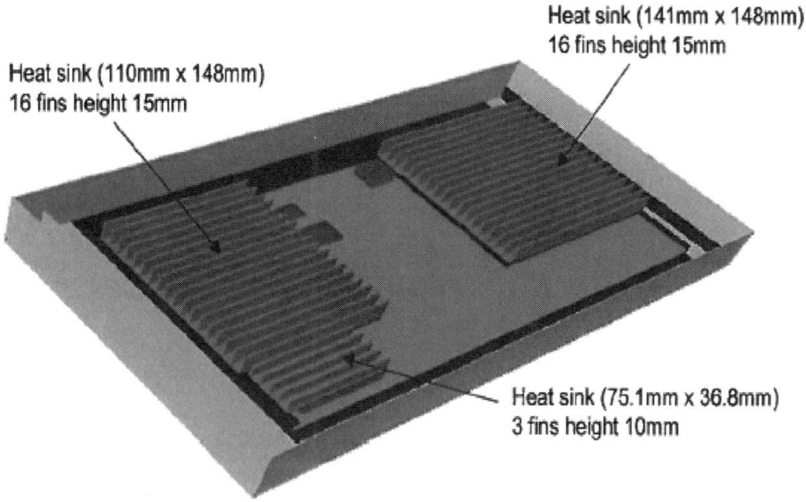

FIGURE 7.7 Various heat sinks on board in box.

Instead of using small heat sinks for individual components which are the general engineering practice, large heat sinks are employed in the present design. Each heat sink covers a group of components with an exception of the 1/4 brick power supply. Heat sinks are made from aluminum with thermal conductivity of 240 W/m °C and emissivity of surface coating at 0.95. The thickness of the fins and the spacing between fins are 2 mm and 6.174 mm, respectively (Figure 7.7).

The temperature distribution over the top surface of the box is shown in Figure 7.8. The maximum temperature on the top of the box is 75.2°C. The component

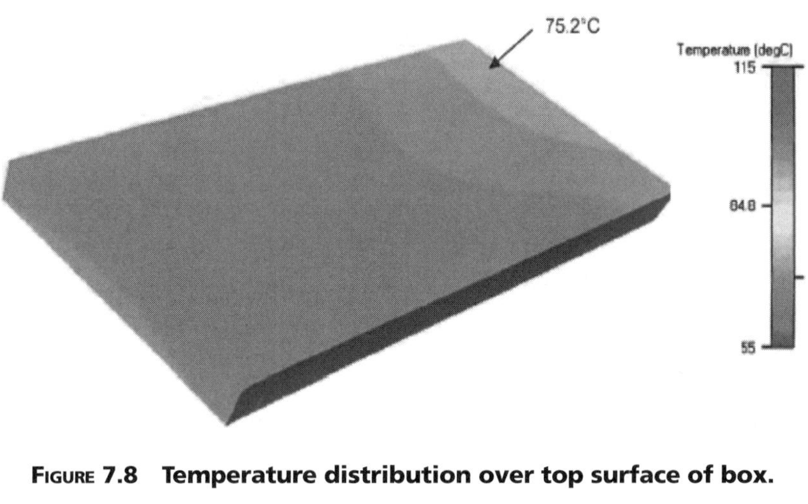

FIGURE 7.8 Temperature distribution over top surface of box.

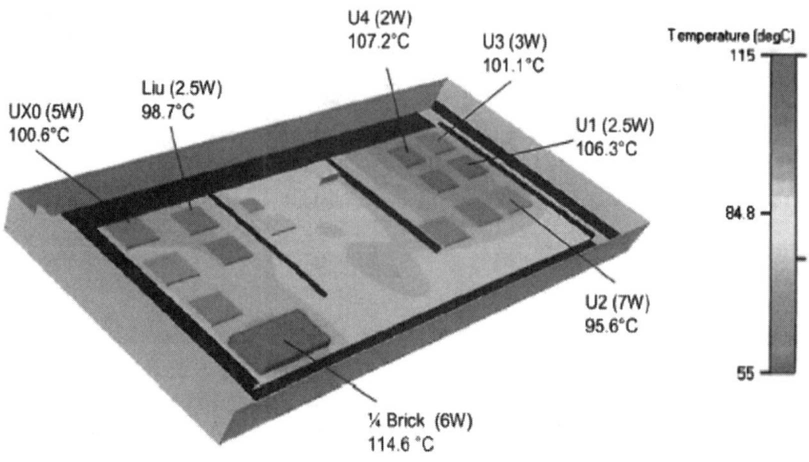

FIGURE 7.9 Component temperatures on board in box.

temperatures are given in Figure 7.9. All components have met the respective temperature limits. To reduce the internal air temperature, a series of small holes on the top surface of the box can be made so that the hot air can vent out the box. To insure a good EMI protection, the opening on the perforate plate must be less than 60%.

7.2.3 Other Applications

The above analysis, which is typical practice in the telecommunication industry, is based on the assumption that the box is suspended in a large open air space. In other words, there is no impact from the surrounding structures on system thermal performance, which represents an ideal condition. It is of interest in examining the thermal performance of the box under the real world conditions.

Table Top Deployment

Though the box is typically mounted to a rack as shown in Figure 7.5; however, it is also often to just place the box on a table. To simulate this condition, a steel plate with the size of 920.6 mm × 399.3 mm × 1 mm is added to the thermal model. It is assumed that the box is in perfect contact with the table top. In order to see the maximum effect of the plate, the assumption of the zero thermal resistance at the interface between the plate and box is made. Since the plate is much larger than the bottom surface of the box, it results in both positive and negative impacts on the box thermal performance. First, the plate acts as a heat sink to enhance the heat transfer from the bottom surface of the box to the ambient. On the other hand, the plate completely blocks the cold airflow from the space below to the box.

Figure 7.10 shows the temperature for the selective components for the case with the box on a steel table top. The figure also includes the temperatures for the cases with (1) stand alone box and (2) the box on a plastic table top. As can be seen from the figure, the stand alone box has the best thermal performance in all three cases under consideration. Due to low thermal conductivity of the plastic, it limits the conduction spreading along the plastic table top which not only provides the thermal insulation to the box but also blocks the cold airflow from the space below the box. Therefore, it significantly increases the component temperatures inside the box. Based on this analysis, it is recommended that the box be placed on metal table in order to achieve a better thermal performance of the box.

The results from Figure 7.10 reveals that material thermal conductivity of the table top have great effects on the component temperatures inside the box under the passive cooling conditions. However, the material thermal conductivity of the table top becomes irrelevant if the space between the box and table top exists as shown in Figure 7.11. This space can easily be created by adding four or six support legs to the bottom of the box. The results also indicate that the component temperatures inside the box will decrease as the space between the box and table top increases and approaches the results of the stand alone box if the distance from the box to the plate is infinite.

Boxes Stack-Up

Because of small configuration, the box is sometimes placed in a cabinet or desktop. To maximize the available space, boxes are often stacked up to the top of each other. The boxes stack-up do not have any effect thermally on the forced air cooled boxes, but it will have severe impacts on the thermal performance of the

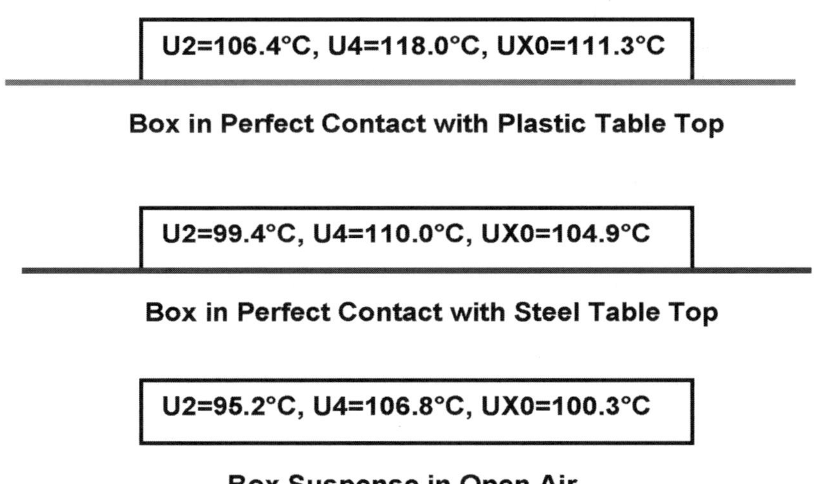

FIGURE 7.10 Component temperatures for boxes with/without plate.

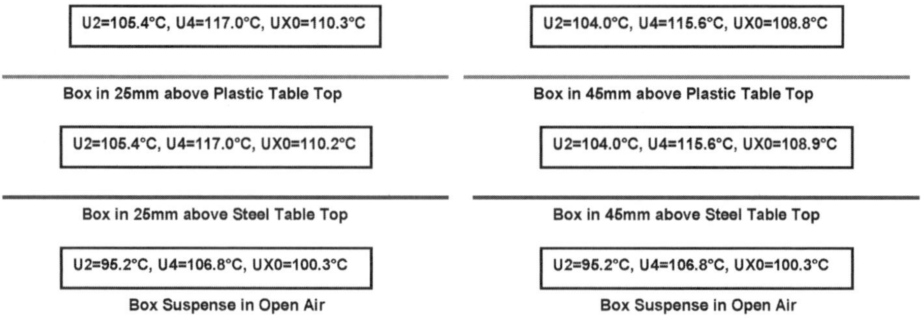

FIGURE 7.11 Component temperatures versus space between box and table top.

boxes under the natural convection environments. The heat transfer capability will almost be reduced by a half for two boxes stack-up. Heat from the top surface of the bottom box is blocked by the bottom surface of the top box. Similarly, heat from the bottom surface of the top box is impeded by the top surface of the bottom box.

Following the discussion in the previous sections, the analysis is extended to the cases with the stack-up boxes on the steel table top, and the results are given in Figure 7.12. For two boxes stack up, the top box is superior to the bottom box thermally. This is because the cold airflow is impeded by the large plate. In addition, there is radiation heat transfer from the top surface of the top box directly to the surrounding cold environments. The similar trend (the top box is coolest) is found in the case of the three boxes stack up on the steel table top. The results also

FIGURE 7.12 Component temperature in single and multiple boxes stack-up on steel table top.

indicate that the middle box is the hottest one among the three boxes. In general, the second box from the top one is hottest for the multiple boxes stack up.

The example presented here is to illustrate the expected trends. The actual numerical values of the component temperatures depend on the system power and configuration as well as component layout. In general, the component temperature increases in an ascending order from the bottom box with an exception of the top box.

The analysis is also extended to study the boxes stack up in an open space and the results are given in Figure 7.13. The component temperatures at the bottom box are about the same or slightly cooler than those of the top box for the current box configuration. Again, the second box from the top box is hottest one in multiple stack-up boxes.

In addition, the effects of the gap between the boxes on the component temperatures are also examined Figure 7.14 shows the effects of the gap size between boxes on the component temperatures. As should be expected, the component temperature decreases as the gap size increasing. In theory, the component temperatures in all boxes will approach to the values of the single box if the gap size is very large (to the infinity).

Due to the cost of floor space in the data center and/or the central office, the facility operators will typically place as many equipment (boxes) into a single rack as possible. Therefore, the configurations presented in Figures 7.13 and 7.14 are much closer to the real world conditions and should be included in the thermal design processes.

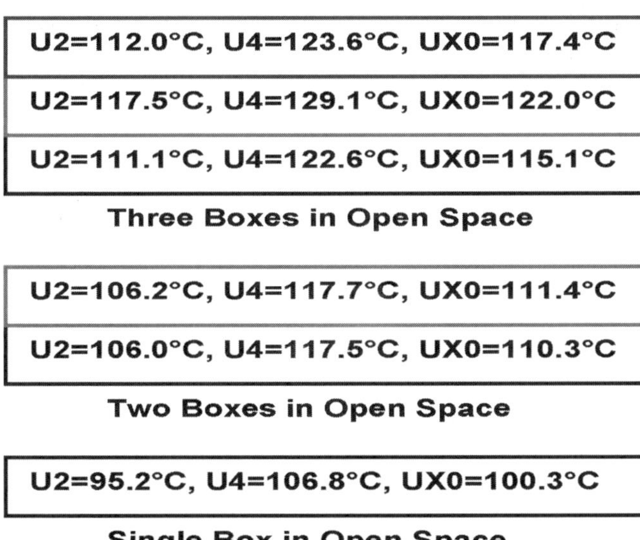

FIGURE 7.13 Component temperatures in single and three boxes stack-up.

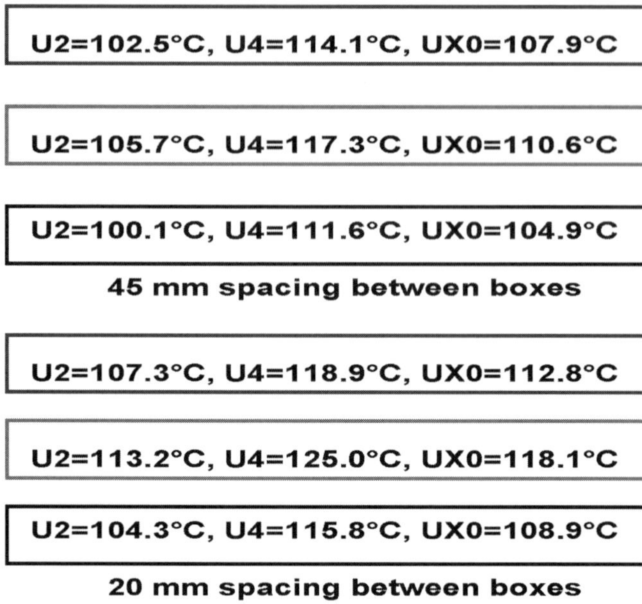

FIGURE 7.14 Effects of gap size on component temperatures in boxes.

REFERENCES

1. Flotherm, a product of Mentor Graphic.
2. Yeh, L. T. and Chu, R. C., Thermal Management of Microelectronic Equipment, ASME Press, 2002.

Chapter 8

OUTDOOR ELECTRONIC EQUIPMENT

The outdoor electronic equipment includes a large variety of various types of systems. Among them are the outside plant (OSP) and the remote radio head (RRH) for telecommunication applications. The OSP refers to all of the physical cabling and supporting infrastructure such as conduit, cabinets, towers or poles, and any associated hardware located between a demarcation point in a switching facility and a demarcation point in another switching center or customer premises. On the other hand, the RRH describes an operator radio control panel that connects to a remote radio transceiver via electrical or wireless interface. From the methods of equipment mounting, the outdoor system can generally be classified into two groups. One is the ground or vehicle mounted equipment and another is the tower or pole mounted systems.

8.1 GROUND MOUNTED OUTDOOR ELECTRONIC EQUIPMENT

For all air-cooled outdoor electronic equipment, heat generated from electronics inside the enclosure is carried away by the air circulated internally. Heat is then transferred to the external stream of the ambient air through an air-to-air heat exchanger and is finally dumped into the ambient.

8.1.1 Thermal Designs of Outdoor Electronic Equipment

All outdoor equipments are subjected to extreme weather conditions, high humidity, high ambient temperature and solar radiation. At the same time, they are expected to operate with high reliability. A large variety of cooling techniques are available in cooling of the outdoor electronics, including telecommunications equipment. These include passive and active forced air cooling, the air conditioners or heat pumps, thermosyphons, and phase change materials (PCMs).

The following case studies are used as examples to describe thermal design methodology for general outdoor electronic equipment:

All electronics are enclosed in a large sealed box [1] with the overall dimensions of 66 in (length) × 36 in (width) × 28 in (depth). The system as shown in Figure 8.1 includes 21 power supplies (116 W) and three racks (269.4 W) (also

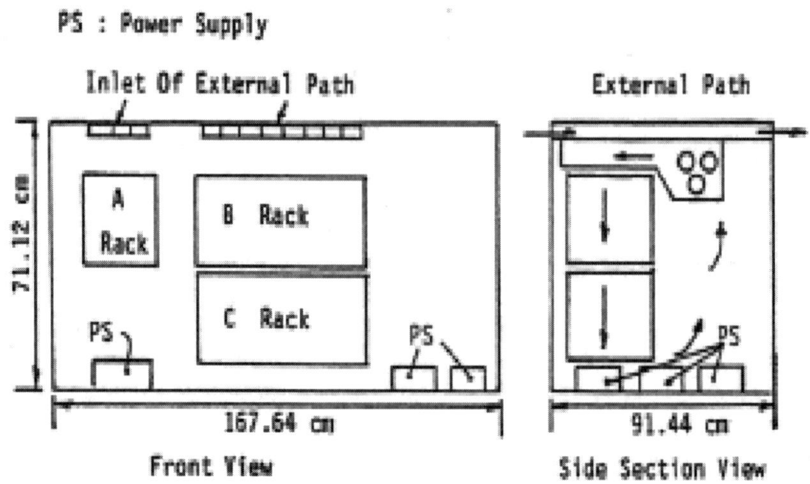

FIGURE 8.1 Sketch of system under consideration.

referred to as card cages). The total system power is 385.4 W. Racks A, B, and C contain 6, 18, and 15 printed circuit boards, respectively. The electronics and the power supply are cooled by the internal air circulated inside the box. The hot internal air is then cooled by the ambient air through an air-to-air heat exchangers as illustrated in Figure 8.2. The internal air is first blown into the air duct by three fans, and is then turned 90° horizontally flowing into the two separate heat exchangers towards Rack A and B/C racks, respectively. After leaving the heat exchangers, the internal air makes another 90° turn downward and enters Rack A and Racks B/C in series, respectively. The air leaving A and C racks flow over the floor from the front to the rear of the box, and then both air streams then combine together, and finally flow sideways toward the internal fans. The hot internal air is used to cool the power supply before it is recirculated into the internal air duct by the fans. Three exhaust fans at the rear of the box draw the ambient air from the front to the rear of the box.

The goal of the thermal design is to meet the following temperature limits:

a. Component junction temperature less than 105°C
b. Power supply base plate temperature less than 85°C

This outdoor equipment is designed for the worst high temperature environment described in Reference 2 as given in Figure 8.3. The calculated values given in the figure are based on the equations developed by Yeh [2]. The figure gives a daily cycle of the ambient air temperature and the direct solar flux on the vertical and horizontal surfaces over 24 hours. The maximum air temperature is 120°F (49°F) at 4 PM and the maximum solar incident is 355 Btu/hr-ft^2 (1120 W/m^2) at a horizontal surface between 12 and 1 PM. The solar absorptivity and emissivity of the surface of the box (equipment) are 0.66 and 0.89, respectively. The actual heat absorption at the surface is the solar incidence time the solar absorptivity.

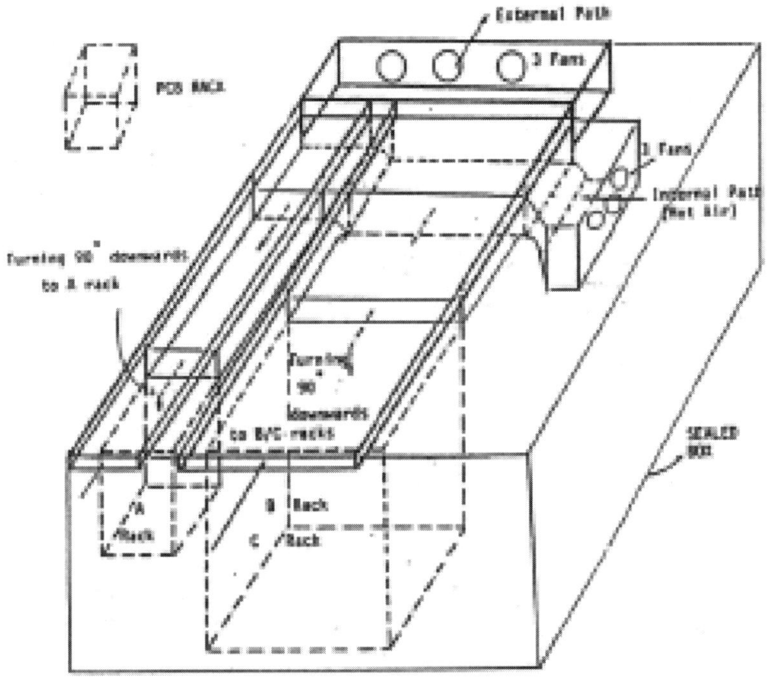

FIGURE 8.2 Internal and external air flow paths.

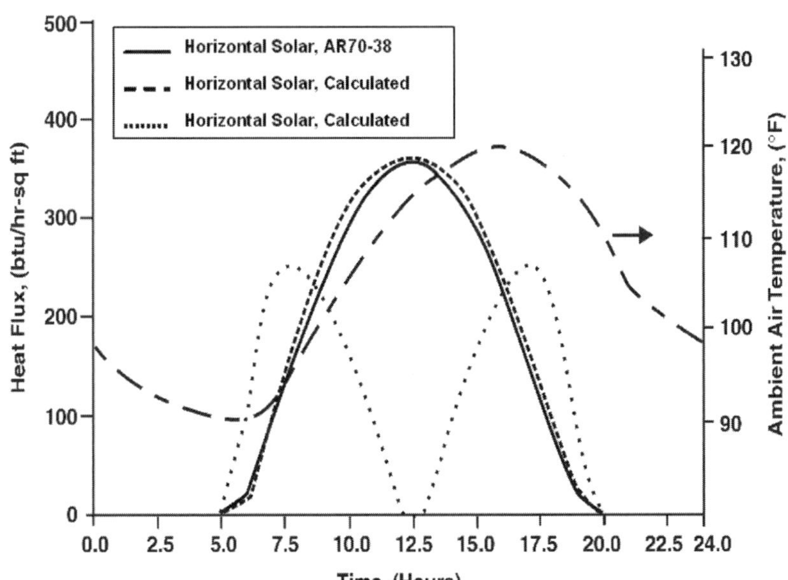

FIGURE 8.3 Hot dry daily cycle of air temperature and solar flux.

A transient thermal analysis is performed on the system over a 24-hour period. The internal air temperature histories at various locations, including at the inlet to Racks A, B, and C are computed as presented in Figure 8.4 and the maximum air temperature to the card cages is 60.2°C at 4 PM. The airflow rates to Rack A and Racks B/C are 113.6 and 290.7 CFM, respectively. The calculated air temperatures and the flow rates are then applied to individual racks to compute the junction temperature of the components under the steady state condition. The results indicate that all components and the power supply have met the respective temperature limits (105°C for microelectronics and 85°C for power supply base plate).

Since the outdoor equipment is exposed to the sun, the solar heat load should always be included in the analysis. The present design approach represents the most realistic conditions by performing the transient analysis first. The maximum air temperature and the flow rate to the boards or box from the transient analysis are then input to the steady state analysis of the boards or box. It should also be noted that the double wall construction with ambient air flowing is adapted for the top wall. The double wall will greatly reduce the solar heating to the electronics inside the box.

The heat pipe heat exchanger as illustrated in Figure 8.5 can be considered to replace the compact heat exchanger. The external and internal air streams are also separated in the heat pipe heat exchanger.

Wankhede et al. [3] evaluated the cooling solutions for various outdoor electronics. Both CFD simulations and thermal tests were performed. Three varieties of outside coating were examined, including white, black, and no coating (plain aluminum finish). The solar absorptivity (α) and emissivity (ε) of three different surface finishes are listed in Table 8.1. Among them, the white oil coating which has the low solar absorptivity and the high surface emissivity makes it a very favorable choice for cooling of the outdoor electronic enclosures.

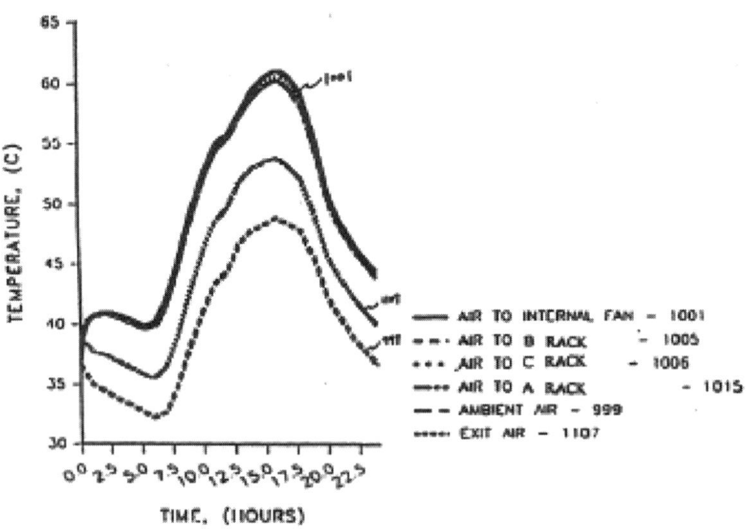

FIGURE 8.4 Internal air temperature history.

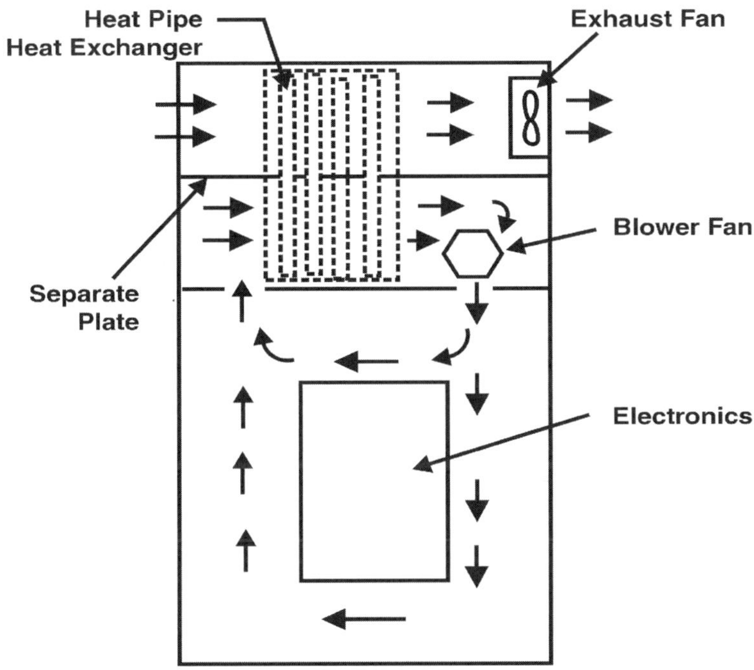

FIGURE 8.5 Heat pipe heat exchanger.

One of configuration under consideration as shown in Figure 8.6 also employs the heat pipe heat exchanger as described in Figure 8.5. The heat exchanger system consists of two parallel heat pipes with the condenser ends located outside of the enclosure. In addition, fins are bonded to the evaporator and the condenser sections of the heat pipes to enhance the heat transfer.

The thermal design utilizes the double wall concept as mentioned in Figure 8.1 to minimize the effect of the solar heating. The schematic of this design concept is illustrated in Figure 8.7. The maximum solar load is assumed to be 600 W/m^2 for Pune city in India during March which is one of the hottest months of the year. The solar heating is applied to three adjacent surfaces, the top and two sides as prescribed by GR-487 [4]. It should be noted that the CFD analysis performed is limited to the steady state conditions. The study concludes that the effect of solar heating on an outdoor enclosure is significant which can increase the internal air temperature by 20%.

Table 8.1 Radiation Properties of Coating

Types of coating	Solar absorptivity (α)	Surface emissivity (ε)
White oil	0.25	0.91
Black coating	0.88	0.88
Plain aluminum	0.08	0.09

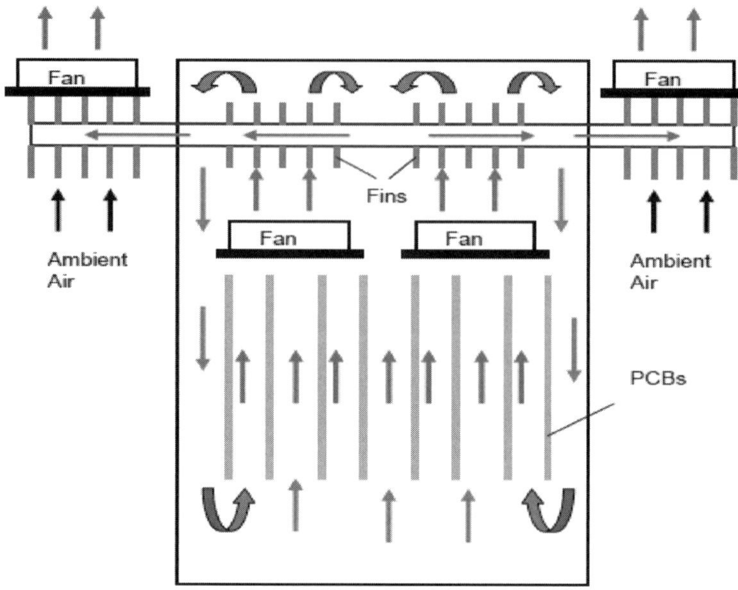

Figure 8.6 Flow patterns inside enclosure.

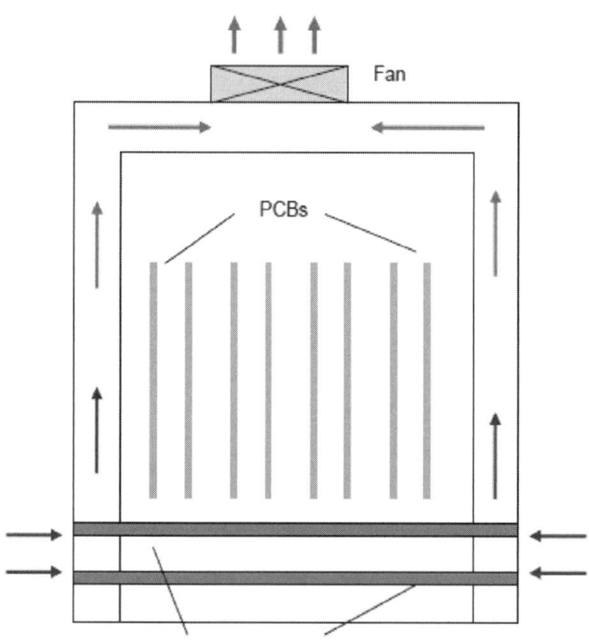

Figure 8.7 Schematic of double wall enclosure.

8.1.2 Thermal Design of OutSide Plant (OSP)

The OSP has received considerable attention in the last few years. This is because of significant increase in the wireless applications and also the great demands of high speed videos and data transmission to the home and offices.

The typical OSP cabinet is shown in Figure 8.8. The OSP includes the wireless and the land line applications; however, it has same basic thermal configuration as illustrated in Figure 8.9.

Outside plants in the past were deployed in the remote locations. However, with the growing consumer demand for higher data/video transmission speeds, the outdoor cabinets are now frequently installed much closer to the users in residential areas. Therefore, the OSP must now operate much quitter. The acoustic noise limit is reduced from 60 to 45 dBA per the newly updated version of GR487 [4]. In addition, the application of the solar heating specifies in GR-487 is different from Example 8.1. The GR-487 describes that solar heating at 750 W/m^2 is applied to any three adjacent surfaces of the equipment (cabinet/enclosure) at the steady state condition.

FIGURE 8.8 External view of typical OSP cabinet.

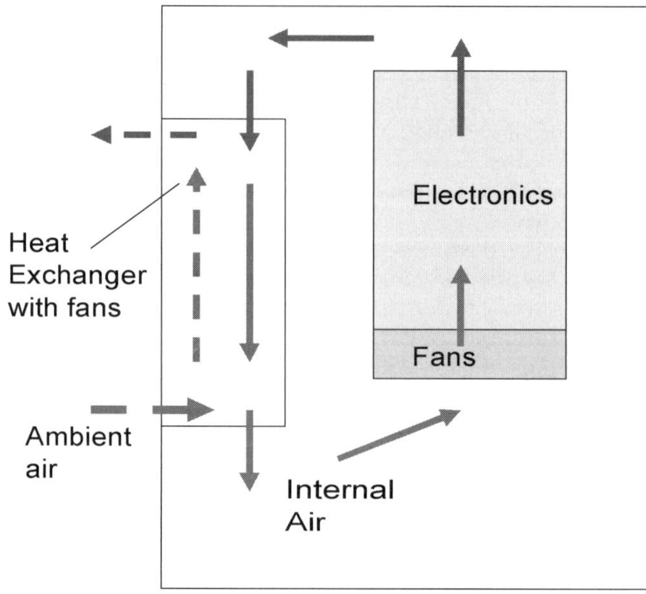

FIGURE 8.9 Basic thermal configurations in outdoor enclosure.

Conventional Heat Exchanger OSP

As shown in Figure 8.9, the OSP generally employs a compact heat exchanger to transfer heat from the cabinet to the ambient. In order to overcome a relatively high-pressure drop across the compact heat exchanger, more powerful fans are required which lead to high acoustic noise and power consumption. The analysis also requires a very large and fine meshed thermal model to represent the compact heat exchanger which leads to a long computational time.

The attempt has been made by Trivedi et al. [5] to develop a simplified compact model for the compact heat exchanger. The cabinet, including one heat exchanger as shown in Figure 8.10 is under consideration. The methodology for development a compact model is to replace the core of the heat exchanger by a volume flow resistance and a volume of source resistance to simulate fluid flow and thermal boundary conditions, respectively. The coefficients of the flow resistance and volume source resistance are obtained from the results of the detailed model. With this simplified thermal model, the mesh count and the computational time have been reduced by 45% and 63%, respectively. The results from all the simulations indicate an excellent agreement between the compact model and the detailed model with the error less than 10%. It should be noted that the solar heating is not included in the analysis.

Mariam [6] provides more detailed information about the thermal design of a typical OSP. The author also extends the analysis to the thermal design of the battery compartment with the thermal electric coolers.

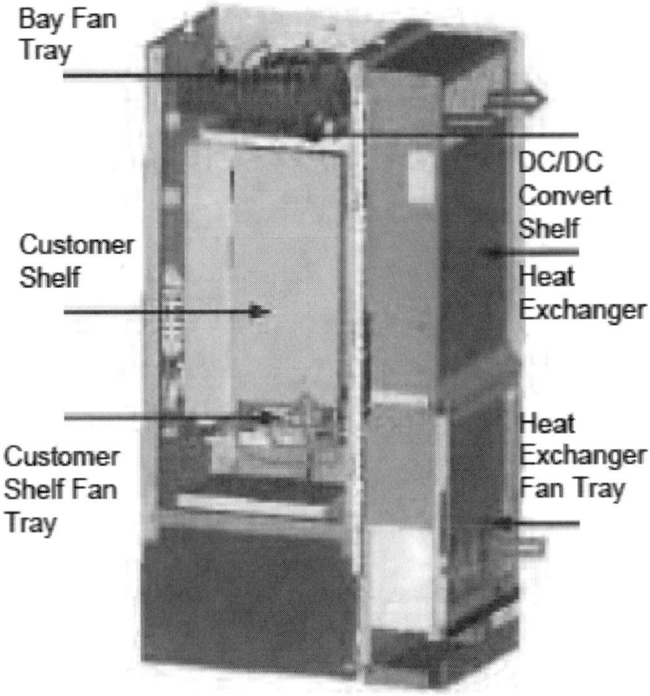

FIGURE 8.10 CAD model of cabinet.

Geo-Thermal Heat Exchanger OSP

To achieve energy efficiency of the equipment, a typical OSP cabinet as shown in Figure 8.8 is placed on the top geothermal heat exchangers as presented in Figure 8.11 [7]. The conventional heat exchanger is replaced by the geothermal heat exchanger.

The cold air from the exit of the geothermal heat exchanger goes upwards from the bottom of the left compartment in the OSP cabinet as shown in Figure 8.11. After flowing over all electronics and the power units, the hot air flows into the right compartment through the duct on the top of the cabinet and is forced down by small fans into the geothermal heat exchanger where the heat is transferred though the heat pipes to the soils. The cooled air once again goes into the left compartment of the OSP cabinet to start the new flow cycle again. In order to enhance the heat transfer, the fins are soldered to both condenser and evaporator sections of the heat pipes. Nickel plating is recommended on external surfaces of the heat pipes to prevent corrosion from contacting with soils. The heat exchanger for the external air stream in Figure 8.9 is basically replaced by U-shaped tubes plus the heat pipes under ground.

The soil temperature depends on the climate, including solar heating and the depth of soil of the specific location. Since the ambient temperature and solar

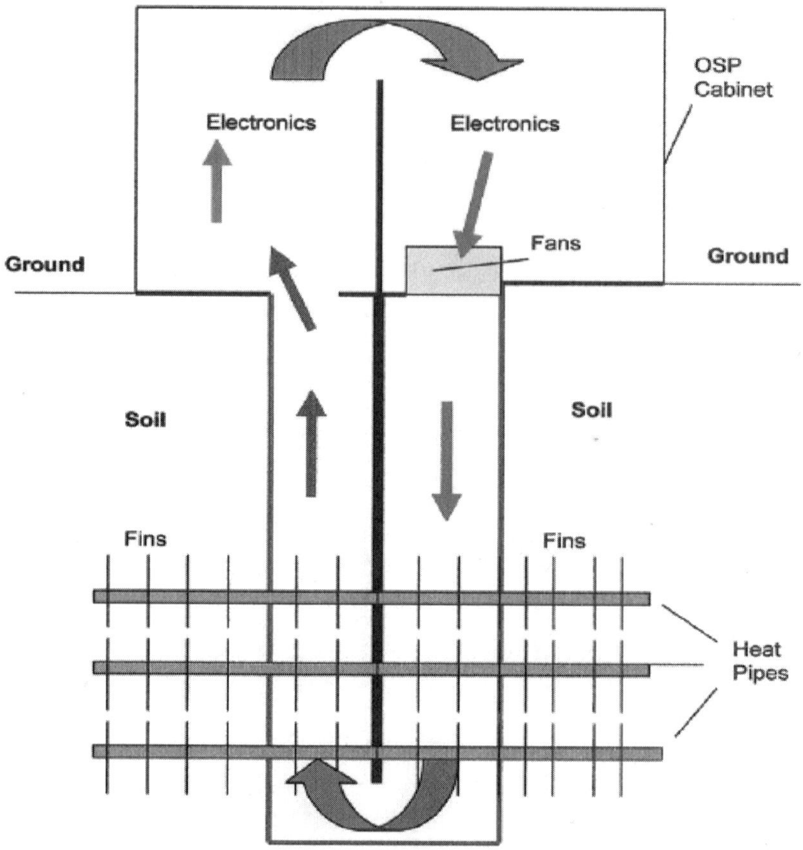

FIGURE 8.11 Geothermal OSP.

heating are function of time, the soil temperature is also a function of time and the depth from the ground surface. The soil temperature can be determined by solving transient one-dimensional heat conduction in a semi-infinite solid with time dependent thermal boundary condition at the ground surface. The soil temperature profiles are illustrated in Figure 8.12. The left figure shows the transient temperature profiles when the ground surface is heated at ($T\infty + \Delta T$) where $T\infty$ is the soil initial temperature from the ground surface to the infinite depth region. On the other hand, the right figure represents the soil temperature profile at a given time with the ground surface at various temperatures, e.g., ($T\infty + \Delta T1$), ($T\infty + \Delta T2$), and ($T\infty + \Delta T3$).

The soil temperature decreases as the depth from the surface of the ground increasing. There are two types of the soil temperature variations. One is due to the daily cycle and another is from the seasonal variation. The soil temperature below 5 ft is generally independent of the daily cycle of the ambient temperature

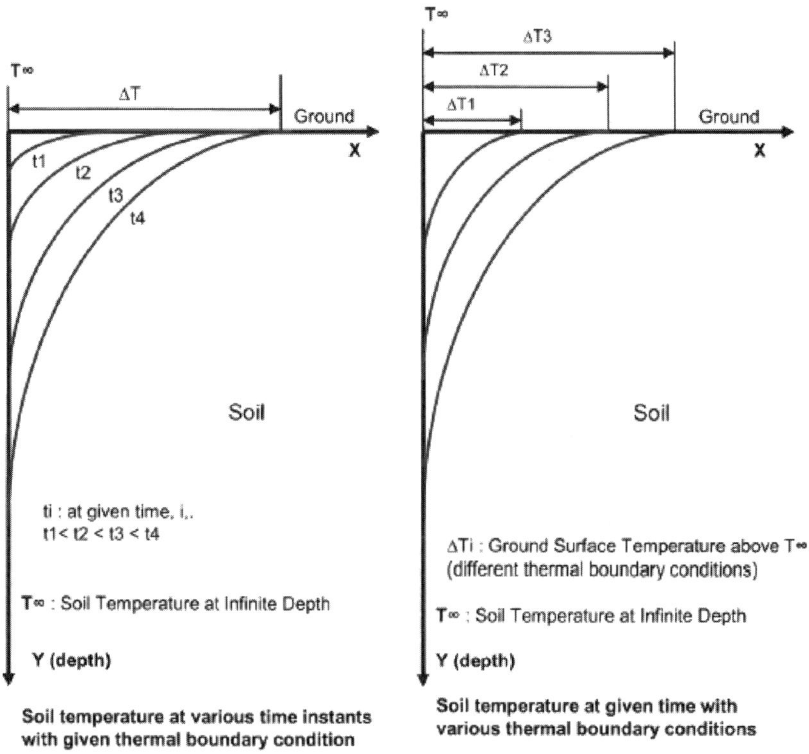

FIGURE 8.12 Typical soil temperature profiles.

and solar heating in the former case. On the other hand, the seasonal soil temperature variation is within about 5°C at about 15 ft below the ground and there is no change with seasons below 50 ft. The soil temperature below the ground surface is typically colder than the ambient air in the summer. Therefore, geothermal OSP is better suitable for the hot climate regions.

The geothermal cabinet is more energy efficient than the conventional one by eliminating the conventional heat exchanger and also using the smaller fans to circulate air through the underground-section. However, there are some inherit issues. The disadvantages of the OSP with geothermal heat exchangers are as follows:

1. Complicated and costly installation procedures
   a. A relatively large hole under the ground is required and the digging may require special permits from the city, state, or federal government agencies
   b. The installation procedure must insure the soils in good contact with the surfaces of the heat pipes for effective heat transfer to the soil
2. Service, repair, and maintenance issues

There is no possible or practical for repairs or services once the equipment is installed under the ground. This is especially critical in the case of natural disasters such as earthquakes.

Other issues are related to the system upgrades. Again, the geothermal heat exchanger cannot be upgraded as needed in the future.

The conventional heat exchangers which are typical compact heat exchangers in general require larger fans in order to overcome the pressure drop over the heat exchanger. Large fans consume more power and also generate higher acoustic noise. The geothermal cabinets can achieve better energy efficiency and lower acoustic noise but have inherit problems as described above. Therefore, the best way to achieve energy efficiency for an outdoor enclosure is to adapt a natural convection for the external stream through the heat exchangers in Figure 8.9. However, due to the ineffective heat transfer of the natural convection, the system power is somewhat limited.

Direct Air Cooled OSP

All previous discussions employ various types of the heat exchangers in the outdoor equipment to separate the internal and the external (ambient air) air streams. This is because no direct contact is permitted between the electronic components and the ambient air for most outdoor equipment. The main reason is to prevent the electronics from contamination by the ambient air. The updated version of GR-487 allows the direct air cooling of the equipment as long as the special air filter which can filter out the dust, sand, soot, pollen, and salt particles is employed.

The direct cooling eliminates the needs of using the heat exchanger. The advantages of the direct air cooling as compared with the heat exchanger cooled OSP are as follows:

a. Operates at the lower temperature due to low supply air temperature,
b. Eliminates the possible hot spots along the internal circulation air path,
c. Increases system reliability and also operation life of electronics,
d. Reduces system complexity and the number of mechanical parts in system,
e. Lower system power consumption resulting in higher COP.

Generally, the direct cooling will lower the internal air temperature of the cabinet by about 10°C as compared with that of the conventional heat exchanger OSP. Gore's ePTFE membrane filter [8] which is naturally waterproof is one of common used filters in the direct air cooling for the outdoor electronic equipment. Comparison of dust protection among various types of the air filter is presented in Figure 8.13 [8]. The results indicate the ePTFE filter is much superior to others.

Garcia and Cosley [9] adapt the double air filters design as given in Figure 8.14 for the outdoor enclosure. The primary one is the typical fibrous air filter for debris and large particulates, and the secondary one is the hydrophobic filter such as ePTFE filter for vapor transmission and small particulates. It is not necessary to use

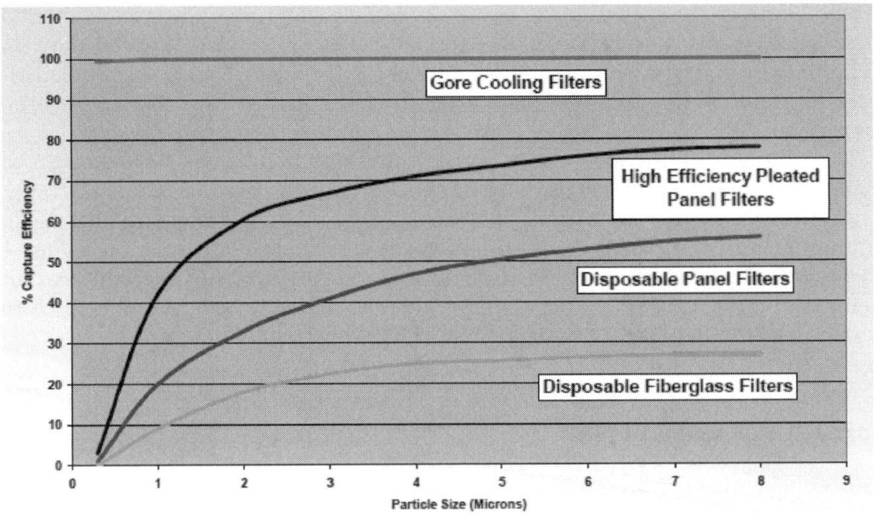

FIGURE 8.13 Dust protection of various types of air filters.

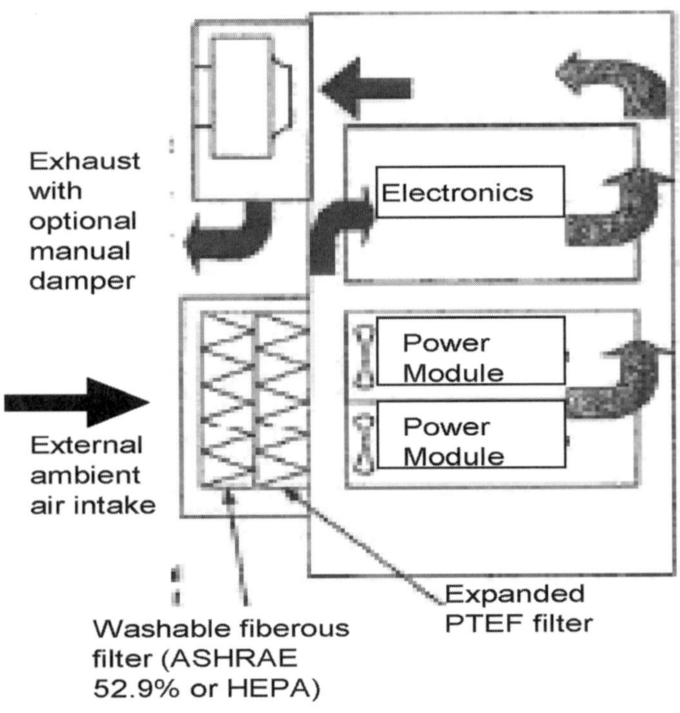

FIGURE 8.14 Filtered vent thermal system concept.

double filters as described in the paper. It depends upon the environmental challenge. The ePTEF filter developed by Gore is designed with layered pre-filtration that will protect from typical ambient dust and debris. If the environmental challenge is more severe, then the use of a separate pre-filter will help to increase the life of the hydrophobic filter.

The results from the field trial tests indicated that the pressure drop across the primary filter increases with the time of exposure during the field trial and, on the other hand, no significant increase in pressure drop over the secondary filter during near a year trial test. The former implies that the primary filter functions properly. This is because the accumulation of dust decreases the available free flow area and increases the pressure drop across the filter. The latter reveals the benefit of the primary filter in prolonging the exposure life of the secondary filter.

Natural Convection OSP

The natural convection has been utilized extensively to cool microelectronic equipment because the passive cooled system is energy-efficient, reliable, and

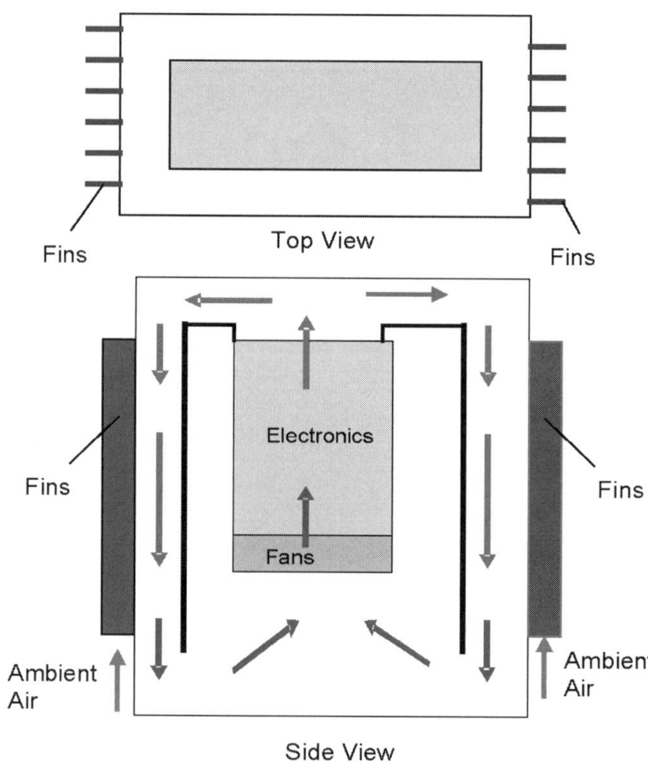

FIGURE 8.15 External natural convection for OSP cabinet.

cost-effective. The natural convection cooled OSP which adopts the natural convection for the external air stream is illustrated in Figure 8.15. As can be seen from the figure, the hot side (internal air stream) is due to forced convection and the cold side (external air stream from the ambient) is by natural convection. By eliminating the fans in the external flow path, the equipment achieves greater values of COP which has the higher energy efficiency and also lowers the acoustic noise. It should be noted that the cabinet sometimes only has one side wall available for heat transfer between the equipment and the ambient. As always, the radiation heat transfer must be included in the passive cooling. It is also often to add a cover or shroud to protect the external fins which forms a counter-flow heat exchanger as shown in Figure 8.16.

Applying the results from his earlier work (10), Yeh estimated the possible cooling capacity of a natural convection OSP. The heat sink configuration and the summary results are presented in Figure 8.17. The heat width is 56" and the height ranges from 26"–60". The heat sink base is assumed at constant temperature of 66°C which is an average value of the inlet and exit air temperature of the heat sink as shown in Figure 8.17. The optimal fin spacing for the heat sink under consideration is 0.513". The heat loss from the heat sink which is based on the correlations [2] for a straight fin array may over-predict over the CFD results by 10%.

The use of the heat exchangers, including the geothermal heat exchangers, as well as natural convection cooling is only capable of providing above ambient temperatures to the electronics inside an OSP. Typical temperature difference between the internal and the ambient air is about 20°C. The direct cooling approach provides a near the ambient temperature to the electronics. By eliminating the heat exchanger, the direct cooling reduces the size of the system thermal control unit and the power consumption as well as the acoustic noise. An air conditioner can be employed to provide the below ambient temperatures to the

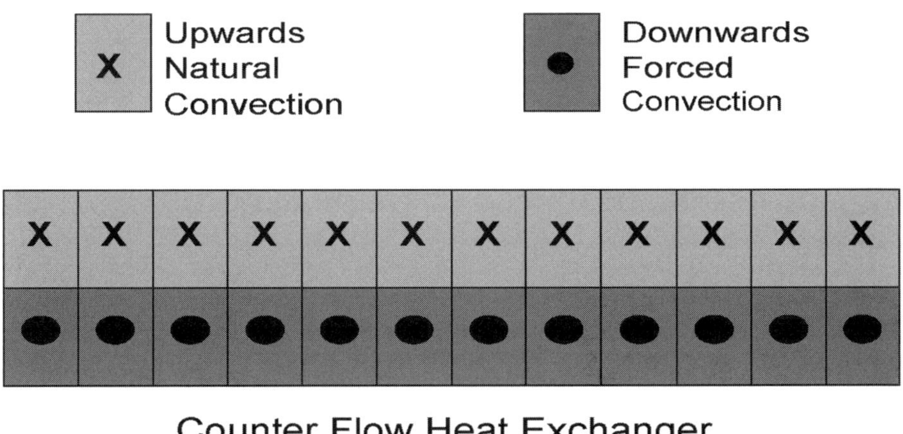

FIGURE 8.16 Counter flow heat exchanger for OSP in passive cooling.

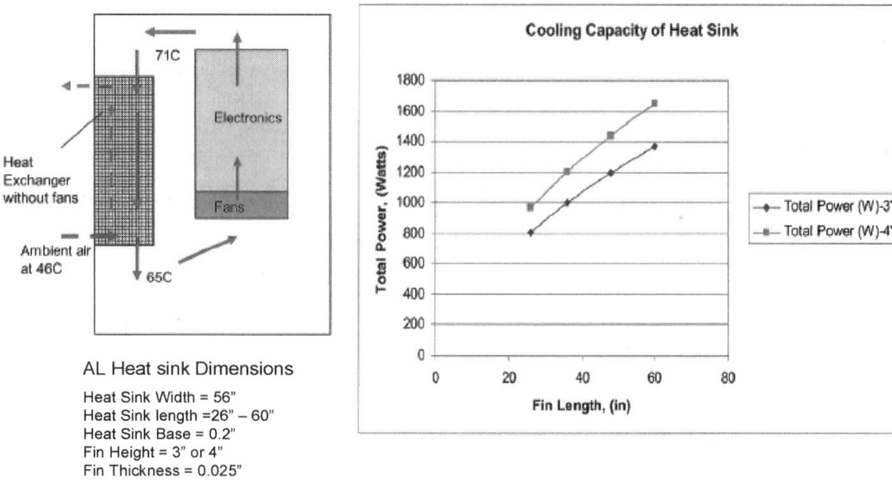

FIGURE 8.17 Estimated cooling capacity of natural convection OSP.

electronics. However, the size increase in the thermal system due to the air conditioner unit results in significant increase in the power consumptions. Another way to provide the below ambient temperatures is to utilize the thermoelectric coolers which require a significant amount of additional (electrical) power input.

Recently, the fuel cell technology, which is one of the green technologies, has received a great interest by the data center industry to improve the energy efficiency. The application of the fuel cell technology has also been extended to the OSP. The current application of this technology is limited to as the back up power (i.e., back up to the batteries in the cabinet). The fuel cell OSP cabinet is currently available in the Europe market. Further utilization of the fuel cell technology to the OSP is expected to increase in the near future.

The summary of various types of cooling technologies employed by OSP is listed in Table 8.2 where the comparisons are made for various types of the OSP

8.2 TOWER/POLE MOUNTED OUTDOOR ELECTRONIC EQUIPMENT

Remote radio heads (RRHs) have become one of the most important subsystems of today's new distributed base stations. The remote radio head contains the base station's RF circuitry plus analog-to-digital/digital-to-analog converters and up/down converters. RRHs also have operation and management processing capabilities and a standardized optical interface to connect to the rest of the base station. Figure 8.18 shows the tower mounted devices.

Table 8.2 Comparisons of Various Cooling Technologies for OSP

Thermal technology	Advantages	Disadvantages
1. Conventional heat exchanger, Figure 8.9	1. Separated internal and external air paths 2. No environmental contamination to electronics 3. Applicable to wide range environments (region/weather)	1. Higher component temperatures due to higher inlet air temperature to electronics 2. High pressure drop over external stream heat exchanger which requires large powerful fans
2. Heat pipe heat exchanger, Figure 8.5	1. Separated internal and external air paths 2. No environmental contamination to electronics 3. Applicable to wide range environments (region/weather)	1. Higher component temperatures due to higher inlet air temperature to electronics 2. Slightly low pressure drop over external stream heat exchanger as compared with Case 1
3. Geothermal heat exchanger, Figure 8.11	1. Separated internal and external air paths 2. No environmental contamination to electronics 3. Few fans needed for underground external air flow path with small pressure drop 4. Less power required and lower acoustic noise 5. Inlet temperature to electronics less dependent at ambient conditions	1. Limited applicable environments (only in hot climatic regions) 2. Soil thermal properties varying with locations 3. Current design for power less than 1000W 4. Complicate snd costly installation processes 5. Possible; additional governmental regulations
4. Direct air cooling, Figure 8.13	1. Lower component temperature due to direct cooling by ambient air 2. No eternal air path heat exchanger (less power required)	1. Expensive air filter needed to prevent environmental contamination on electronics 2. Some limitations to applicable environmental (location/weather conditions)
5. External natural convection cooling, Figure 3.15	1. No fan needed for external air path in heat exchanger 2. Less power required and lower acoustic noise	1. Limited system power due to less effective heat transfer in natural convection

FIGURE 8.18 Picture of tower mounted devices.

8.2.1 Thermal Design of Remote Radio Head

The weight and volume are the key design factors for any tower or pole mounted equipment in addition to the thermal requirement. A passive cooling, which is simple, reliable, and cost-effective, is most often employed for these equipments. A typical remote radio head (RRH) with a heat sink is shown in Figure 8.19. RRHs are for the wireless applications and are typically mounted on the pole, tower or even the top of the roof. As any outdoor equipment, the RRH is subjected to the solar radiation heating and the design and operation are also governed by GR-487.

FIGURE 8.19 Typical remote radio head.

Natural convection from vertical finned heat sinks has been studied extensively in the past. Among them, Yeh et al. [10] performed CFD analysis of the finned heat sink under the natural convection environments. The authors first optimize the fin spacing and also examine the thermal performance of various fin configurations such as continuous fins, staggered fins, and in-line fins as shown in Figure 8.20 and found that the thermal performance of the continuous finned heat sink is the best. Yeh [11] further extends the analysis to study the effect of the cover or shroud over the finned heat sink. The cover will enhance the natural convection heat transfer. In addition, based on the limited results, there is no effect of cover on the heat transfer if the distance between the heat sink fin tip and the cover is greater than 2.5 times of the fin height.

The RRH typically employs the extruded finned heat sinks as shown in the above figure in the past because of easy manufacturing and low cost. However, the extrusion heat sink is bulky and heavy due to the thick base and fins which also limit the available heat transfer surface area. The heat sink is typical made from aluminum. The size of the extrusion heat sink will be significantly increased in order to support the increasing system power. Since in most of cases, the heat load is not uniformly distributed over the heat sink base, it would require a thick base to spread heat to the entire heat sink.

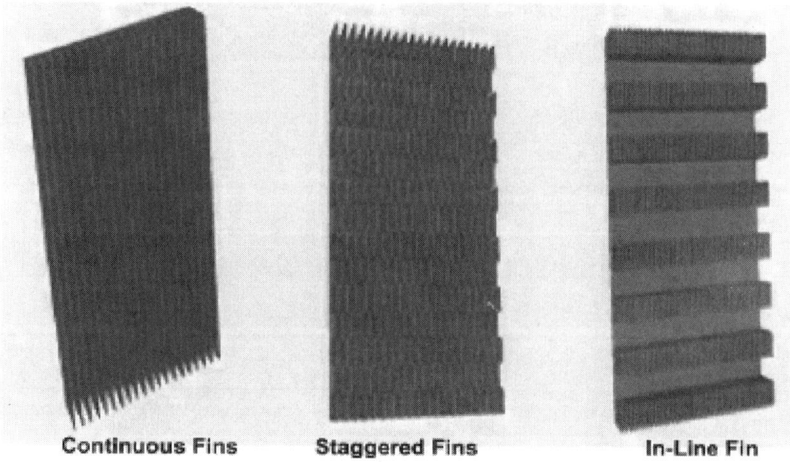

FIGURE 8.20 Various types of finned heat sinks.

Skandakumaran et al. [12] adapt a thin, flat two-phase active heat spreader (AHS) as shown in Figure 8.21 to improve the thermal performance of the tower mounted RRH heat sinks. The dimensions of the heat sink base are 280 mm × 540 mm. The active heat spreader (AHS) is basically a vapor chamber device. The improvement is limited to the base plate of the power amplifier. The experiments and simulations are performed using four different extrusion heat sink configurations as presented in the figure. The thermal conductivity of an extruded aluminum heat sink is 187 W/m °C. The heat sinks are tested at 75 W of thermal power under the natural convection environment. The test and simulation results (maximum

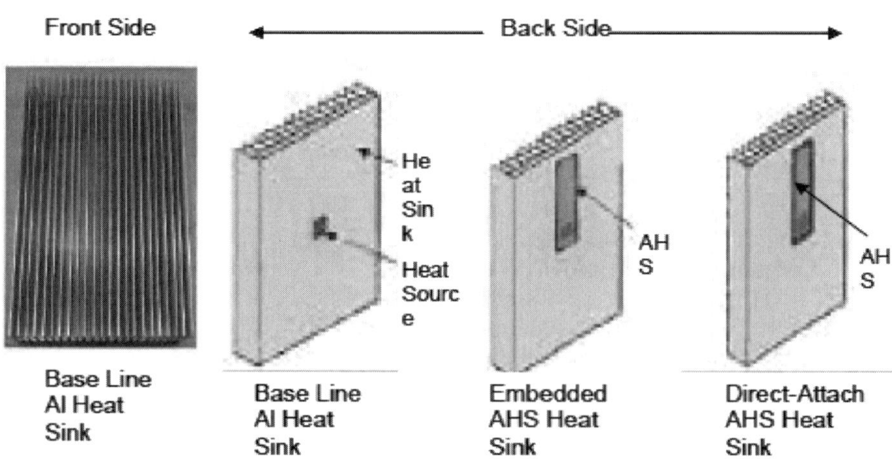

FIGURE 8.21 Active heat spreader heat sinks.

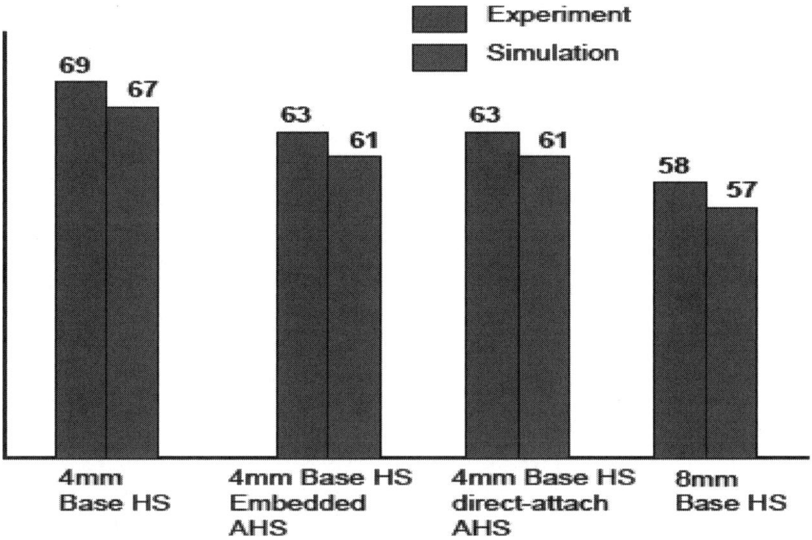

FIGURE 8.22 Extruded heat sink thermal solutions.

temperatures) are given in Figure 8.22. The results reveal that the 4-mm base heat sink exhibits the highest temperature while the 8mm base heat sink has the lowest temperature. It appears that AHS device only improve the heat transfer by about 10% due to the small size of the device which limits the heat spreading capability.

Though copper has much higher thermal conductivity, the copper heat sinks are also eliminated for consideration because of the weight factor. The conventional heat pipes can be embedded into the base of heat sink to enhance heat spreading. However, it may lead to significantly increase the thickness of the heat sink base. In addition, a great number of heat pipes are required for a large heat sink. Another option is to adopt embedded graphite in the heat sink base. The thermal conductivities for the graphite are 1000+ W/m °C in the *XY* plane and 20 W/m °C in the *Z* plane (thickness direction). The density of the graphite is about the same value as the aluminum.

Liquid has been employed to cool the high power supercomputers recently. Similarly, liquid can be considered for the RRH to deal with the increased system power. Thermosyphon is often employed for the passive cooling conditions. Thermosyphon includes two different cooling modes as illustrated in Figure 8.23. One is the closed tube and another is closed-loop thermosyphones. The former is a self-contained closed tube which is similar to the heat pipe, and the latter is two-phase closed loop system. The important difference between a heat pipe and a thermosyphon is at the mechanism of condensate return in these devices. A heat pipe relies on the capillary action of an internal wick and a working fluid, whereas a thermosyphon employs an external force such as gravity or centrifugal force for condensate return to the evaporator. Since there is no wick involved, a closed tube thermosyphon is often referred to a wickless heat pipe.

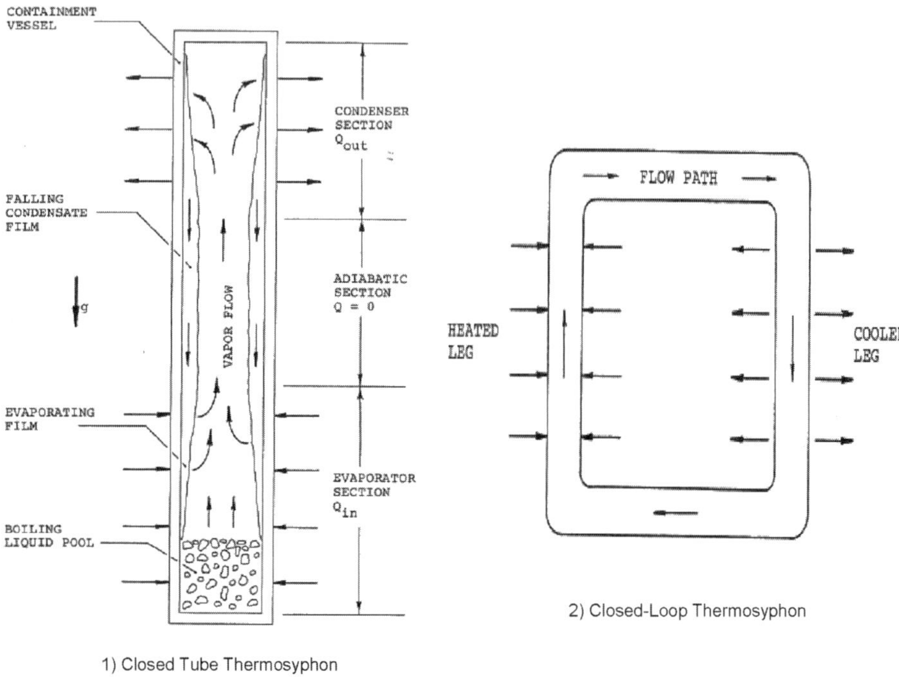

FIGURE 8.23 Basic thermosyphone models.

Similar to the heat pipes, the closed tube thermosyphons can be embedded into the base of a finned heat sink. The cooling scheme of a closed-loop thermosyphon for RRH is illustrated in Figure 8.24. The section of the heated leg is a liquid cold plate in a direct contact with the RRH power module, and the section of the cooled leg is the base of the finned heat sink. The vapor flow is generated due to the heating from the RRH electronics in the hot leg and vapor is then condensed to liquid by rejecting heat to the ambient through the finned heat sink by natural convection and radiation. The flow circulation as shown in the figure is because of the difference in fluid densities in the hot and the cold legs.

Since it is for outdoor applications, the freezing point of the fluid must be below −40°C. The dielectric fluids such as Fluorinert liquids (FC-43, -75, -77, -87) from 3M company and Ethylene Glycol/water mixture. The disadvantage of the thermosyphon is that the thermal performance of the system will degrade significantly as the inclined angle from the vertical plane increasing, and it practically ceases to perform at the horizontal position.

The extruded heat sink is bulky and heavy because of thick fins which are typically greater than 1.5 mm. With the advances in the manufacturing technologies, the plain fin heat sink can be made to replace the extruded heat sink in order to reduce its weight. For example, for fin height of 2 in., the fin thickness and the base can be reduced to 0.02 in. and 0.1 in., respectively. The weight for a plain fin

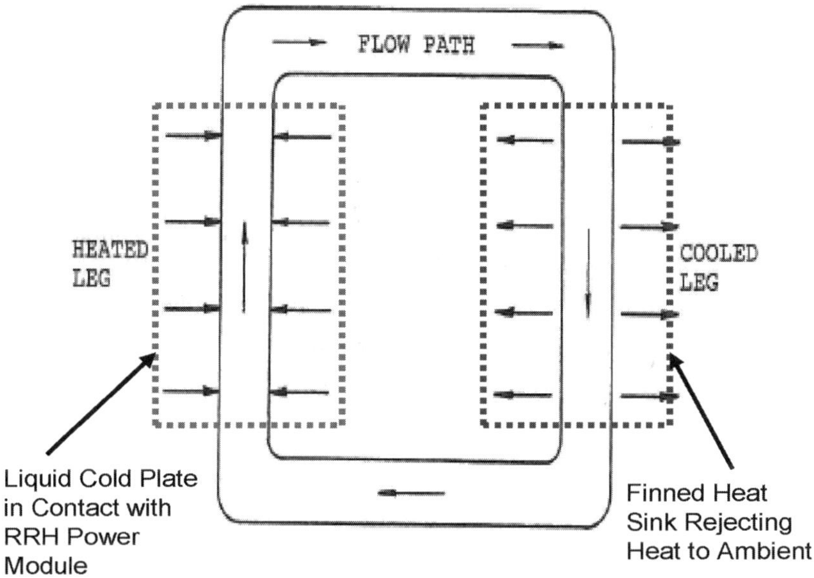

FIGURE 8.24 Passive cooling with closed-loop thermosyphon.

heat sink with the overall dimensions of 10.4 in. (width) × 12 in. (length in flow direction) is only about 1 kg. This lightweight heat sink can be deplored at any orientation, even at the horizontal position.

Another popular tower or pole mounted outdoor equipment is the ONT (optical network terminal) as shown in Figure 8.25 which are subjected to the direct solar heating and the ambient of 46°C. However, some of the ONTs are installed inside of the garage of the home where the equipment is not directly exposed to the sun but the air inside the garage is hotter than the ambient air due to the solar heating on the roof and the sidewalls of the building. Therefore, the equipment must be designed for the minimum "ambient" temperature of 65°C under the natural convection conditions.

To reduce the cost, Corning plastic outdoor enclosure is frequently employed. The basic thermal solution for an ONT can be illustrated in Figure 8.26. Generally, the electronic components are enclosed in a metal box to prevent EMI leakage to the environment and also from a direct contact between the ambient air and the components. The system power for a fully functioned ONT could be up to 25 W, and the overall dimensions of the envelope are approximately 300 mm (height) × 240 mm (width) × 80 mm (depth). The solar absorptivity of the plastic is about 0.44.

Solar heating is a major concern for any outdoor equipment. The solar thermal loads in some cases are greater than the heat load from electronics inside the equipment. Since the infrared (IR) energy emitted by the Sun is at much shorter wavelengths than those emitted by a body near room temperature. This

FIGURE 8.25 A typical ONT with cover open.

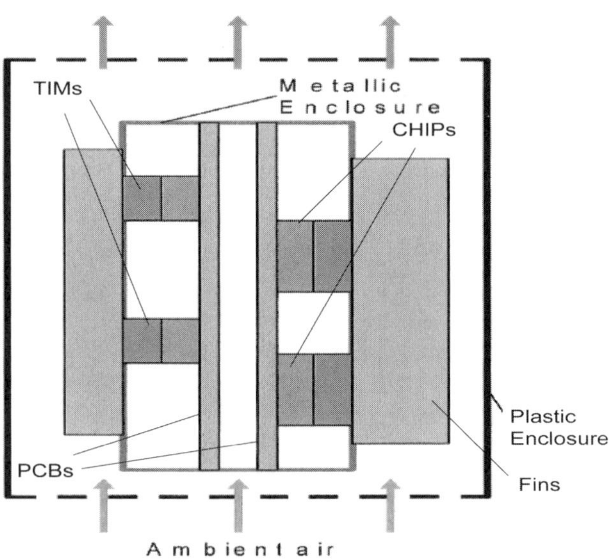

FIGURE 8.26 Basic configuration of ONT.

distinct characteristic allows for the some of thermal control coatings with low solar absorptivity and high emissivity being applied to the external surfaces of the system for thermal management of solar radiation heating. This special coating (AXJ-4020) with the solar absorptivity of 0.15 and the material emissivity of 0.9 is currently available from AZ Technology.

REFERENCES

1. Yeh, L. T., "Transient Thermal Analysis of Outdoor Electronic Equipment", EEP-Vol. 23/HTD-Vol. 356, CAE/CAD and Thermal Management Issues in Electronic Systems, ASME, 1997.
2. Yeh, L. T., and Chu, R. C., Thermal Management of Microelectronic Equipment, ASME Press, NY, 2002.
3. Wankhede, M., Khaire, V., Goswami, A., and Mahajan, S. D., "Evaluation of Cooling Solutions for Outdoor Electronics", Volume 16, No. 3, Electronics Cooling, 2011.
4. Telcordia Technologies Generic Requirements, GR-487-CORE, Issue 3, April, 2009.
5. Trivedi, A., Agonafer, D., Sivanandan, D., Hendrix, M., and Sahrapour, A., "Compact Modeling of A Telecommunication Cabient", Proceedings of IMECE2008, Boston, MA, October 31–November 6, 2008.
6. Mariam, F. A. I., Thermal Management of Outside Plant Telecommunication Cabinet: Design and CFD Modeling Methodology, MS Thesis, The University of Texas at Arlington, May, 2010.
7. Hong, Yuping, Shengqin, J., Liqian, Z., Qiao, C., Bianco, C., Griffa, G., and Cucchietti, C., "Cooling System of Outdoor Cabinet using Underground Heat Pipe", INTELEC, San Diego, CA, 2008.
8. Braswell, M., and Poon, W., "High Reliability with Direct Air Cooling", IWPC Workshop, Chicago, IL, July 30, 2007.
9. Garcia, M. P., and Cosley, M. R., "Ambient Air Cooling of Electronics in An Outdoor Environment", 26th International Telecommunications Energy Conference (INTELEC), September 19–23, 2004.
10. Yeh, L. T., Yeh, J., and Chung, B. T. F., "Natural Convection from Finned Heat Sinks", Proceedings of IPACK2007, Vancouver, British Columbia, Canada, July 8–12, 2007.
11. Yeh, L. T., "Natural Convection from Finned Heat Sinks with/without Cover/Shroud", 19th International Symposium on Transport Phenomena, Reykjavik, Iceland, August 17th–21st 2008.
12. Skandakumaran, P., Khanikar, V., Smalc, M., Norley, J., and Reis, B., "Passive, Lightweight Thermal Solutions for Remote Radio Head (RRH) Electronics", Itherm, 2008, 2008.

Chapter 9

LIQUID COOLED SYSTEMS

In the cooling of electronic equipment, liquid cooling is frequently applied to high power electronic equipment because of its high heat transfer capability. Liquid cooling can further be divided into single- and two-phase flow systems. The latter involves phase change processes such as boiling or condensation that greatly increases the heat transfer capability by utilization of the latent heat. In addition, the phase change process takes place at the constant temperature.

Since there is no commercially available liquid cooled telecommunication equipment presently, this chapter first focuses at the fundamental theories of both single- and two-phase flow heat transfer, follows with the coolant selection, and then briefly describes current liquid cooling technologies as well as their applications in the electronics industry in general. Finally, system thermal design will be discussed.

9.1 CONVECTION AND BOILING HEAT TRANSFER

Basic fundamentals of single phase convection and boiling heat transfer will be discussed in this section.

9.1.1 Single-Phase Forced Convection

The heat transfer coefficient of a single-phase flow under forced convection is a function of the Reynolds number (*Re*), Prandtl number (*Pr*), and the thermal boundary conditions as well as the shape of the tube for fully developed laminar flow or for fluids with Prandtl number less than one. The Prandtl number is defined as $Pr = \upsilon/\alpha$, where υ is kinematic viscosity and α is thermal diffusivity of the fluids. It is well known that the heat transfer coefficient increases with the Reynolds number ($\rho VL/\mu$), where ρ is density of fluid, V is velocity, L is the characteristic length, and μ is viscosity of fluid.

The current cooling technologies used in the telecommunication industry are still limited to the air cooling. The major difference between the air cooling and the liquid cooling is due to the effects of the Prandtl number of the fluids. The Nusselt number or heat transfer coefficient is in general as a function of $Pr^{1/3}$. For most of gases, the Prandtl number ranges from 0.7 to 1.0 which basically eliminates the effect of the Prandtl number on the heat transfer coefficient. The Prandtl number also has a great effect on the boundary layer development which can be illustrated in Figure 9.1, where δ and δ_{th} are the thickness of the hydrodynamic and thermal boundary layers, respectively. The heat and momentum are diffused through the fluid at the same rate if the Prandtl is unit. If the Prandtl number is greater than 1, the velocity profile will develop more rapidly than the temperature profile, whereas the opposite trends are found in a fluid with Prandtl number less than 1.

It should be noted that the temperature profile starts to develop at the point where the heat transfer between the fluid and wall actually takes place. In other words, due to heat conduction along the heating wall, the thermal boundary layer will start at the upstream of the actually heated location. If the Prandtl number is greater than 5, the velocity profile generally will develop so much faster than the temperature profile that a solution based on a fully developed velocity profile will apply quite accurately, even without accounting for the hydrodynamic starting length. Such a condition frequently prevails in the dielectric fluids that are widely used for electronic equipment because the Prandtl number for most dielectric fluids is in general greater than 20 at the room temperatures. On the other hand, for the case with Prandtl number approaches zero, the velocity profile never actually develops while the temperature profile does develop. A uniform velocity profile (slug flow) can be assumed over the entire thermal boundary layer.

In laminar flow or fluids with a Prandtl number less than 1, the heat transfer coefficient depends not only on the thermal boundary conditions but also the shape of cross section of the tube as shown in Table 9.1 [1] The two basic thermal conditions are constant axial wall temperature (*T* condition) and constant heat flux (H condition). For symmetrically heated noncircular tubes, two common cases are:

1. H1 condition: constant axial wall heat flux with uniform peripheral wall temperature (variable peripheral heat flux).
2. H2 condition: constant axial and peripheral wall heat fluxes (variable peripheral wall temperature).

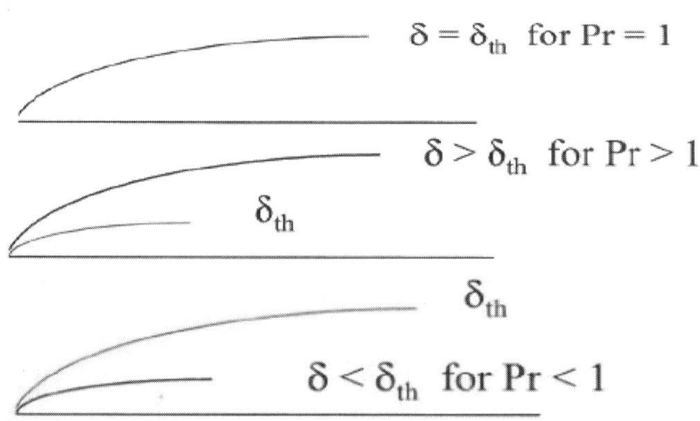

FIGURE 9.1 Prandtl number effect on boundary layers.

Table 9.1 Nusselt Number and Friction Factor in Fully Developed Laminar Flow

Tube Shape	f • Re	Nu_T	Nu_{H1}	Nu_{H2}
Rectangular Tube: Aspect Ratio (α) = 2a/2b 2a: tube height; 2b: tube width				
1.0	14.227	2.976	3.608	3.091
0.9	14.261	–	3.620	–
0.8	14.378	–	3.664	–
0.714	14.565	3.077	3.734	–
0.667	14.712	3.117	3.790	–
0.6	14.980	–	3.895	–
0.5	15.548	3.391	4.123	3.020
0.4	16.368	–	4.472	–
0.333	17.090	3.956	4.795	2.970
0.3	17.512	–	4.990	–
0.25	18.233	4.439	5.331	2.940
0.2	19.071	–	5.738	2.930
0.167	19.702	5.137	6.049	2.930
0.125	20.585	5.597	6.490	2.940
0.1	21.169	–	6.785	2.950
0.05	22.477	–	7.451	–
0	24.000	7.541	8.235	8.235
Circular tube	16.000	3.657	4.364	4.364
Triangle tube (equilateral)	13.333	2.470	3.110	–

It should be noted all above discussions are applicable for either liquid or gas flow.

In addition to the heat transfer coefficient, the friction factor is also needed in computing the pressure drop in a system. The general relation for pressure drop in a duct can be expressed as follows:

$$\Delta P_t = \Delta P_f + \Delta P_s + \Delta P_a \pm \Delta P_g \tag{9.1}$$

where

ΔP_t = total pressure drop,
ΔP_f = pressure drop due to friction
$= 4f(L/D_h)(G^2/2\rho_m g_c)$,
ΔP_s = pressure drop due to additional losses such as entrance, exit, bends, fittings, valves, etc.
$= \Sigma K(G^2/2\rho_m g_c)$,
ΔP_a = pressure drop due to flow acceleration (or momentum changes)
$= (1/\rho_2 - 1/\rho_1)(G^2/2g_c)$,

ΔP_g = pressure drop or gain due to gravity in a vertical flow
= $\rho_m gH/g_c$
"+" for vertical upward flow
"–" for vertical downward flow,
D_h = hydraulic diameter,
F = Fanning friction factor,
G = Mass velocity
= ρV (V is flow velocity),
g = acceleration,
g_c = gravitational constant
= 32.1739 lbm-ft/lbf-sec^2 = 4.17×10^8 lbm-ft/lbf-hr^2,
H = tube vertical height,
K = loss coefficient

9.1.2 Boiling Heat Transfer

From the viewpoint of thermodynamics, when liquid is heated at a constant pressure, the temperature and volume of liquid will increase from A to B as shown in Figure 9.2 [1]. Beyond point B, any additional heat added to the system is to change liquid phase to vapor phase, and Point B is the saturated liquid at a given pressure. In the region between points B and C, liquid and vapor phases coexist in equilibrium. The amount of heat required to convert a unit of mass from liquid to vapor is called the latent heat of vaporization, h_{fg} (= $h_g - h_f$), where h_g and h_f are the enthalpy of saturated vapor and liquid, respectively. In this region, the addition of heat increases the vapor fraction and the volume of mixture, however, the temperature of the two-phase mixtures still remains at the saturated temperature. At point C, all the liquid has vaporized and is referred to as the saturated vapor. Further heating will result in a temperature rise of vapor (superheated vapor) along the constant pressure line.

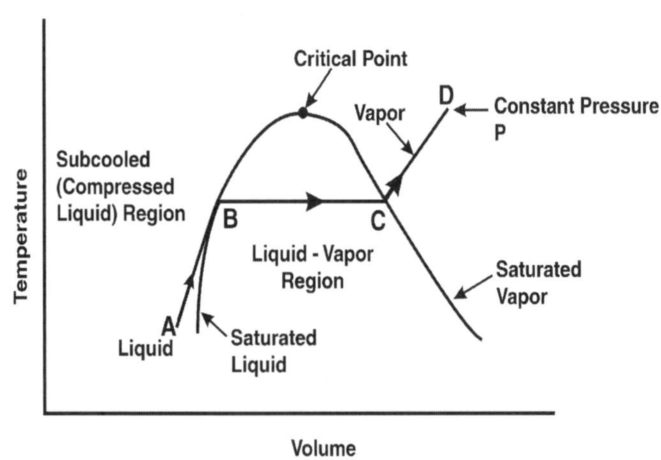

FIGURE 9.2 Liquid at constant pressure heating process.

Boiling is the process of evaporation associated with vapor bubbles in liquid. The change of phase from liquid to vapor is caused by heat transfer or pressure changes. The two basic types of boiling are pool boiling and flow boiling. The former is the boiling on heated surfaces submerged in a pool of stationary liquid, while the latter is the boiling on a surface in a flowing stream of liquid. Depending on the liquid temperature, boiling can also be classified into subcooled boiling, in which the bulk fluid temperature is below the saturation temperature, and saturated boiling, in which the bulk fluid temperature is uniformly at the saturation temperature. The boiling curves for both pool and flow boiling are illustrated in Figure 9.3, and their trends are very similar.

The heat transfer is so effective in the nucleate boiling region that the temperature difference between the heating wall and the liquid is very small, typically less than 5°C. As the heat flux is increased continuously, the number of nucleation sites on the heating surface increases until the maximum heat flux is reached at point C. The maximum heat flux is often called the critical heat flux (CHF). At point C, the vapor generation rate is so high that the surface is blanketed with a vapor film which prevents liquid from contacting the surface.

For pool boiling, in Region CD as given in Figure 9.3, a portion of the heated surface will undergo film boiling while others are under nucleate boiling, therefore, the vapor film is unstable. A smooth transition from nucleate to film boiling can only occur under temperature controlled conditions, which can be accomplished by condensing vapors or by convection from high temperature fluids with high mass flow rates. However, on the other hand, with electric heating, it is impossible to operate in region CD. When the power is increased, the operating point shifts rapidly from point C to C′, which is at a much higher temperature. A physical burnout of material could occur if point C′ is above the melting point of the solid surface. If the temperature of point C′ is below the melting point, operations can be maintained along the curve DE. The region DE is called film boiling because a layer of vapor film continuously covers the surface and liquid does not contact the surface. Because of high surface temperature in this region, radiation becomes an important factor. The operating condition will change directly from point D to nucleate boiling at point D′ if the heat flux is reduced below point D. Then, power again must be increased to point C before the process can shift to film boiling.

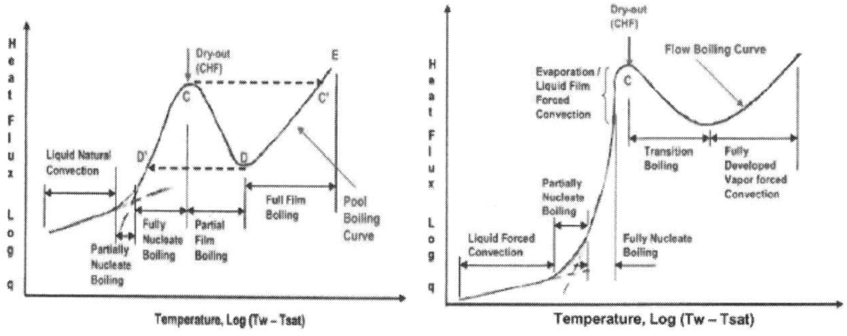

FIGURE 9.3 Boiling curves for pool and flow boiling.

9.2 COOLANT SELECTION

Selection of a coolant for a liquid cooled system is a very important design consideration. A number of factors must be considered when selecting a coolant for a given system. The first step is to establish the important system characteristics, such as weight/volume limitation, power consumption requirement, and leak-tightness. In general, a fluid may be very good in some respects and quite deficient in others. Therefore, there is no single, ideal fluid that is good for all systems and applications. Consequently, various fluid properties of the coolants should be evaluated against the particular systems and applications under consideration. The important properties that affect the selection of a coolant were discussed in details by Knight [2] and these properties include toxicity, flammability, flash point, vapor pressure, pour and freezing points, thermal decomposition, dielectric strength, effects of moisture. One should never overlook the properties which do not have any effects on the system thermal performance. For examples, the toxicity and flammability are critical to the system safety and leakage. The fluid electric strength and the effects of moisture will have great impacts on the system electric insulation requirements.

Coolants must be chemically compatible with the materials, which they will contact within the coolant loop, and from leakage. Also, particular attention must be paid to the selection of seals, gaskets, and adhesives used in the cooling system. Kelly [3] has rated the compatibility of a number of common compounds and some of the general classes of fluids that might be used as coolants. For cooling of electronics, the dielectric property of a liquid is also very important, especially for a direct liquid cooling. The reader is encouraged to read both references.

The fluid properties that are directly related to thermal and hydrodynamic performance of the coolant are specific heat (c_p), thermal conductivity (k), viscosity (μ), and density (ρ). Specific heat is an indicator of the coolant's ability to store thermal energy within a given mass. Thus, it is desired that the fluid have a high specific heat. Thermal conductivity defines the coolant's ability to transfer heat and should be high. Fluid viscosity increases the fluid resistance, which results in a high system pressure drop and required high pumping power, and also reduces the effectiveness of heat transfer, therefore, the viscosity of the fluid should be as low as possible. The density of the coolant relates to the mass of the fluids. The higher the density, the more mass is in a given volume. The density, unlike specific heat and thermal conductivity, also has some negative attributes. Fluid weight and pumping power increase with the density. Thus, it may be desirable to have either a large or a small density, depending on the specific application.

The important properties affecting the boiling process for a given fluid are surface tension (σ) and latent heat of vaporization (h_{fg}). A coolant with a low value of surface tension will result in a small contact angle and wet the heat transfer surface. This high wetting nature of the fluid deactivates many potential nucleation sites on the surface that can lead to a significant surface temperature overshoot at boiling incipience. Care must also be exercised because the liquids having a low surface tension are more likely to cause leakage problems by sipping through seals, gaskets, cracks, connectors, etc.

The coolants may be rated on the basis of heat transfer characteristics as they affect mass flow rate, pumping power, volume, and weight. From the viewpoint of thermo-hydraulic performance, fluids with a high heat transfer coefficient always have a higher flow resistance that requires a higher pumping power to overcome the higher system pressure drop. Depending on the system requirement, a careful balance between the heat transfer coefficient and pressure drop is needed to achieve an acceptable design for thermal control of the system.

For convenience, a figure of merit (FOM) is frequently used to compare the effectiveness of various fluids. The FOM which is a dimensionless parameter represents the relationship among key properties or characteristics of a fluid under prescribed flow conditions and is used to guide the selection and optimization of fluids. FOMs based on heat transfer consideration only are derived from the governing thermal transport relations for single-phase (liquid) and two-phase (vapor and liquid) flow, including forced and free convection; boiling incipience; and critical heat flux for pool and flow boiling [4]. For example, for natural convection, the FOM can be defined as follows:

$$\mathrm{FOM} = [\beta \rho^2 c_p k^{(1/n-1)}/\mu]^n \tag{9.2}$$

where n is 0.25 for laminar flow and 0.33 for turbulent flow.

It is often used in the past to express the FOM to be function of the heat transfer coefficient for the forced convection condition. For a fully developed laminar forced convection, the heat transfer coefficient, h, only depends on the thermal conductivity, k, and the channel hydraulic diameter, d, i.e., $h \propto k/d$. Based on this definition, the FOM does not include the specific heat of fluids in the case of forced convection. Therefore, this definition for FOM appears to be incomplete. Any good heat transfer fluids should possess the ability to transfer heat between the wall and the fluid and should also have the ability to transport heat in the fluid that is related to the specific heat of the fluid. Furthermore, in practical engineering applications, it is beneficial to maximize heat transfer performance while minimizing the required pump power to save the power consumption, system weight and cost. As proposed by Yeh and Chu [1], the FOM for a single-phase flow under forced convection be defined as follows:

$$\mathrm{FOM} = c_p h/P \tag{9.3}$$

where P is the required pump power. If the pump power is not a consideration factor, e.g., natural convection, let $P = 1$ in Equation (9.2).

For fully developed flow, Equation (9.3) can be simplified as follows

$$\mathrm{FOM}_L\ \alpha(k c_p \rho/\mu) \text{ for laminar flow} \tag{9.4a}$$

and

$$\mathrm{FOM}_T\ \alpha(k^{0.67} c_p^{\,1.33} \rho^{1.05}/\mu^{0.72}) \text{ for turbulent flow.} \tag{9.4b}$$

Table 9.2 Coolant Figures of Merit at 40°C

Fluid	Laminar	Turbulent
Water	14.207	27.688
FC-43	0.155	0.633
FC-75	0.405	1.249
FC-78	0.666	1.751
Coolanol-25	0.250	0.980
Coolanol-45	0.071	0.397
Glycol/water	1.382	4.482
PAO(2CST)	0.222	0.950

Notes:
(1) Constant mass flow rate
(2) Based on English units: k (Btu/hr-ft-°F), c_p (Btu/lbm-°F), ρ(lbm/ft^3), μ (lbm/hr-ft)

Numerical values of FOM for various fluids at 104°F (40°C) are given in Table 9.2 [1]. The table indicates that water is the best coolant and is followed by ethylene glycol/water (62.5% and 37.5% by weight). Since thermal properties of fluids are a function of temperature, the temperature may greatly affect numerical values of FOM. One needs to consider the operation temperature range of the system in determining the FOM. The constants listed in Equations (9.4a) and (9.4b) may vary with the correlations of the heat transfer and friction factor being employed in Equations (9.3).

In general, a dielectric liquid which has a high dielectric strength to prevent electrical breakdown is a preferred choice in cooling of electronics. Unfortunately, the heat transfer capability of the dielectric fluids is much lower than that of the water. The dielectric liquid is not normally required for the cases with indirect liquid cooling. However, it is a must for the immersion cooling where liquid is in a direct contact with electronics. Due to very low freezing point, typically below −40°C, dielectric fluids should be employed in all outdoor electronic equipment. Among dielectric fluids, fluorocarbon liquids (FC-series) are the most commonly employed in cooling of electronics.

In summary, any good liquid coolant should have the following general characteristics:

Physical Properties

a. High fluid density (for heat transfer, low density from weight aspect)
b. Low pour or freezing point and low vapor pressure
c. High coefficient of thermal expansion (for free convection, low from pressure aspect)
d. High specific heat and thermal conductivity
e. High flash point and high self-ignition temperature

Electrical Properties

a. Low dielectric constant
b. High dielectric strength

c. Low dissipation factor
d. High volume resistivity

Functional properties

a. High thermal stability
b. High oxidative stability
c. High hydrolytic stability
d. Good lubricity
e. Inertness to system materials (or compatibility)
f. Non-toxicity
g. Low cost
h. Low contamination (fouling) and good clean up characteristics

9.3 LIQUID COOLING TECHNOLOGIES

9.3.1 Direct Immersion Cooling

The simplest method of cooling is to immerse the electronic equipment in a pool of liquid. Due to the direct contact between the liquid and electronics, a dielectric fluid is a must. In other words, the liquid must have the dielectric strength needed to provide electrical isolation between adjacent power/ground conductors and the signal lines. In addition, the liquid must also be compatible with the materials of all parts in the systems.

The thermal properties of all dielectric fluids are poor as compared with the water. The basic characteristics of the dielectric fluids are as follows:

1. Excellent electrical properties, high dielectric strength, low dielectric constant, and high electrical resistivity
2. High thermal stability
3. Low chemical reactivity
4. Poor thermal properties: low thermal conductivity, specific heat, and heat of vaporization
5. Extremely low surface tension
6. High solubility of air

Liquid immersion remains a primary candidate for cooling of electronics, especially for high power electronic systems. For low heat dissipation systems, natural convection takes place in liquids. With increased power, the heat transfer will shift from natural convection to pool boiling.

Parks and Bergles [5] performed a series of experiments to investigate the effect of the heater size on the heat transfer coefficient in the case of natural convection. For flush heaters, the heat transfer coefficient increases as the heater width decreases. The width effect is more pronounced in R113 than in water. The data indicates that the heat transfer coefficient is more than 20% higher than those predicted by conventional correlations. This is attributed to leading edge effects.

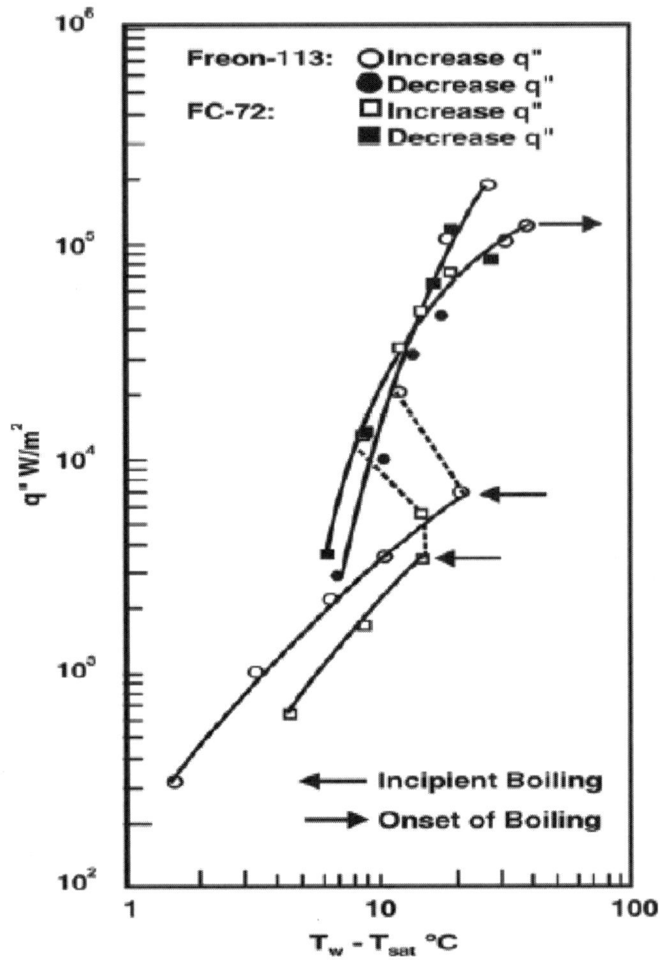

Figure 9.4 Pool boiling data for a plain tube.

Marto and Lepere [6] conducted pool-boiling experiments with R-113 and FC-72 on a plain copper tube with an outside diameter of 15.8 mm and the results are given in Figure 9.4. As can be seen from the figure, a greater incipient boiling superheat is required for R-113 than for FC-72. As stated above, the dielectric fluids have extremely low surface tension which often results in a large temperature overshoot in the case of pool boiling as shown in Figure 9.4. The large temperature overshoot is undesirable and unacceptable for electronic equipment. The methods to eliminate or minimize the temperature overshoot at the incipience of boiling are presented by Yeh and Chu [1].

The liquid-encapsulated module [7] as shown in Figure 9.5 was developed by IBM. Heat generated by the immersed chips in perfluorinated liquid (FC-86) is carried by natural convection to the walls and is then dumped into the ambient

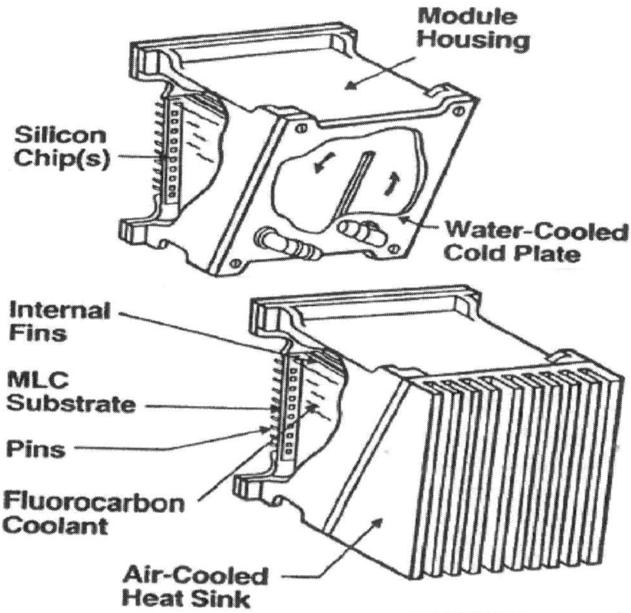

FIGURE 9.5 IBM liquid encapsulated module (LEM).

through the external air-cooled heat sink or to a liquid cold plate. With the saturated temperature of 62°C at the atmospheres pressure, the component junction temperature is at 85°C. In addition, the author also discussed the nature of the departure from natural convection (DNC) and departure from the nuclear boiling (DNB) as illustrated in Figure 9.6. The former deals with the transition from the

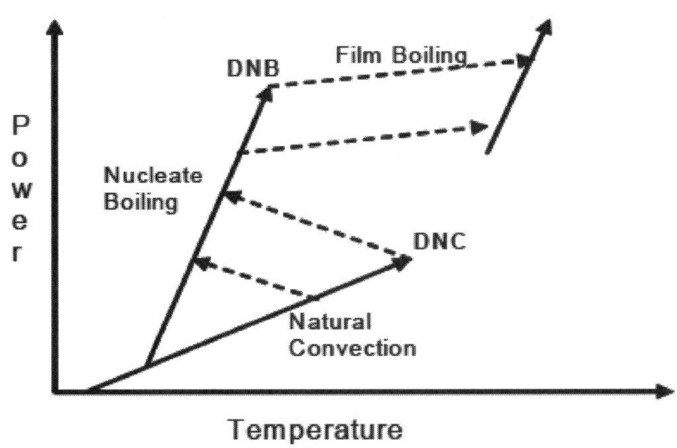

FIGURE 9.6 Boiling curve with DNC and DNB.

natural convection to the nucleate boiling while the latter describes the transition of the heat transfer region from the nucleate boiling to the film boiling. The DNC is generally associated with an excessive temperature over-shoot as shown in Figure 9.4. However, on the other hand, the DNB often results in extremely high wall temperature due to the poor heat transfer in the film boiling.

9.3.2 Indirect Liquid Cooling

Forced convection through a cold plate is frequently used to cool electronic equipment. In most cases, flow is unidirectional from inlet to exit, which leads to a significant temperature gradient in the cold plate. A counter flow arrangement proposed by Yeh [8] as shown in Figure 9.7 may be applied to overcome the large temperature gradient. Typical temperature profiles of the cold plate and fluid for unidirectional flow and counter flow systems are illustrated in Figure 9.8. The advantages of using counter flow over a unidirectional flow system are lowering the cold plate wall temperature and the temperature gradient. Yeh [8] provides an analytical solution given below for the cases where the wall heat dissipation and the heat transfer coefficients are functions of the space.

$$\theta_x = T_2(x) - T_1(x) = \theta_0 e^{-Bx} + Q(x)e^{-Bx} \tag{9.5a}$$

$$\theta_0 = T_2(0) - T_1(0) = \{\{[2h_2(x)W/[h_1(x) + h_2(x)]]\int^L q''(x)dx$$

$$-\{h_1(x)h_2(x)W/[h_1(x) + h_2(x)]\}\int^L \theta_x dx\}/C_2 \tag{9.5b}$$

$$T_1(x) - T_1(0) = \{h_1(x)W/\{C_1[h_1(x) + h_2(x)]\}\}\int^x [2q''(x) + h_2(x)\theta_x]dx \tag{9.5c}$$

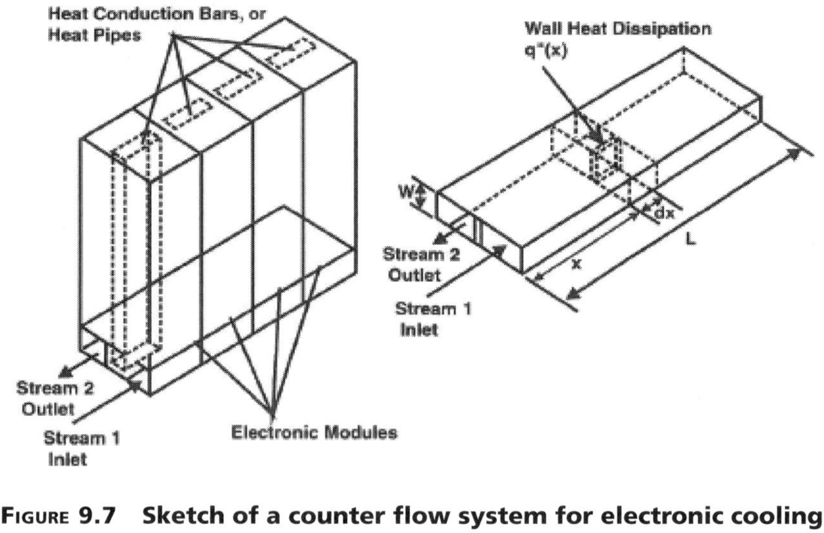

FIGURE 9.7 Sketch of a counter flow system for electronic cooling.

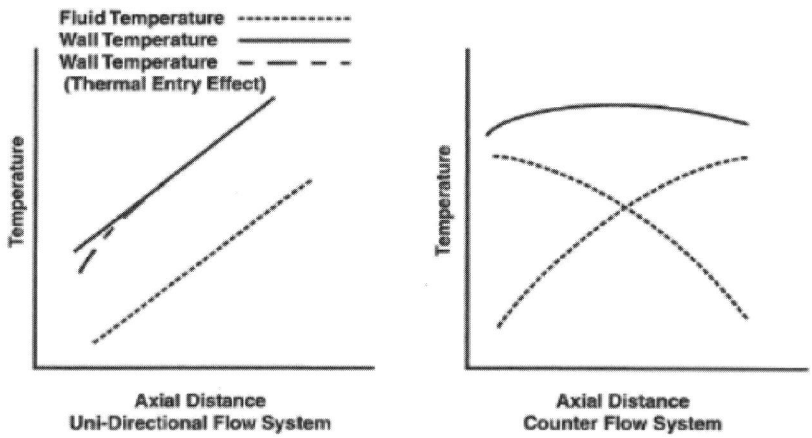

FIGURE 9.8 Cold wall temperature distribution for unidirectional and counter flow.

$$T_w(x) = [2q''(x) + h_1(x)T_1(x) + h_2(x)T_2(x)]/[h_1(x) + h_2(x)] \quad (9.5d)$$

where

$$A - 2W[C_2h_1(x) + C_1h_2(x)]/\{C_1C_2[h_1(x) + h_2(x)]\},$$

$$B = Wh_1(x)h_2(x)(C_2 - C_1)/\{C_1C_2[h_1(x) + h_2(x)]\},$$

$$Q(x) = -\int^x Aq'(x')e^{-Bx'}dx',$$

$$C_i = m_i cp_i,\ i = 1, 2.$$

This approach was further extended to a multiple-channel counter flow system [9, 10] and has been applied to solid state phased array radar [11].

The focus of this chapter is aiming at the mini or micro channel flow which has practical applications for cooling of microelectronics. A liquid cooling of chip package based on the microchannel technology as given in Figure 9.9 was first proposed by Tuckerman [12, 13]. The integrated microchannels are directly at the back of the silicon chip which greatly reduces the thermal path length from the heat sources to the coolant. The height and width of individual cooling channels are 300 μm and 50 μm, respectively. The power of 790 W is dissipated over the 1 cm × 1 cm chip. The experimental results with the water flow rate of 8.6 ml/sec indicate that the temperature difference from the coolant inlet and maximum cold plate temperature is 71°C and the calculated heat transfer coefficient is 11.13 W/cm² °C. Thermally, the microchannel cooling is very effective but its pressure drop which is 31 psi (214 kPa) at the flow rate of 8.6 ml/sec is also too high because of the small cooling channels to be acceptable for practical applications.

Yeh [14] performed similar tests with various types of cold plates. Figure 9.10 shows the picture of one of tested cold plates (the brass cold plate). The overall

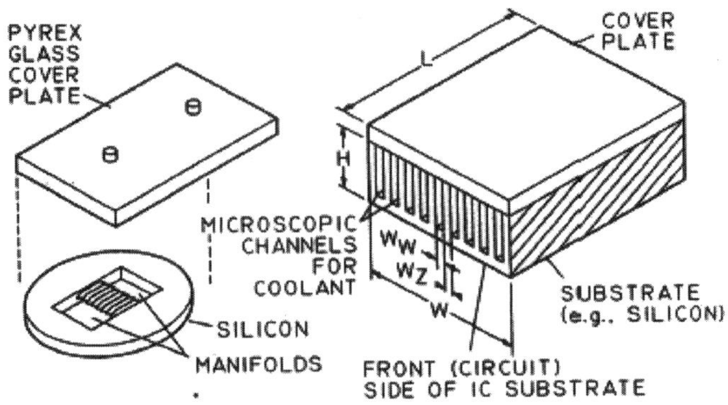

FIGURE 9.9 Microchannel cooled silicon chip.

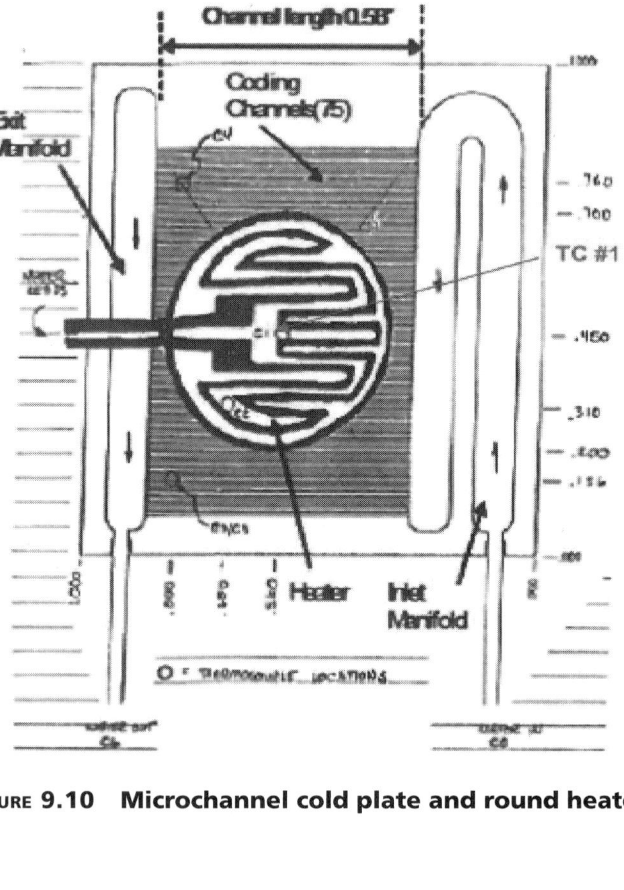

FIGURE 9.10 Microchannel cold plate and round heater.

dimensions of the cold plate are 1 in. × 1 in. × 0.064 in. The cold plate has 75 channels with the cross section of each channel is 0.01 in. (0.254 mm) by 0.006 in. (0.1524 mm). A round foil heater with the cross section area of 0.15 in.^{2} is also shown in the figure. Four thermocouples were instrumented on the backside of the cold plate (opposite to the heater side).

Since no thermocouple is installed at the heater side of the cold plate, to determine the temperature on the heater side of the cold plate, a thermal model was set up to simulate the test conditions. The measured and predicted temperatures at the center of cooling section are presented in Figure 9.11. As can be seen from the figure, an excellent agreement is found between the test and simulation results (average about 1°C difference). The simulation also predicts the maximum cold plate temperature on the heater side which is at the center of the heater. The predicted maximum cold plate temperature at this location with the flow rate of water at 33 ml/min is 40°C above the inlet water temperature (21.8°) when the heater dissipates 50 W with the equivalent heat flux of 334 W/ in.^{2} (518 W/cm^2). It should be noted the flow rates used in Tuckerman and Yeh tests are 8.6 ml/sec and 33 ml/min, respectively.

A thermal test was also performed on a swirl cold plate that has 30 trenches as shown in Figure 9.12. The cross section of each trench is 0.01 in. by 0.016 in. Water enters the center of the cold plate which impinges the heated side of the cold plate and flows radically outwards along the channels to the outer ring as indicated in Figure 9.12. The overall dimensions of this aluminum cold plate are same as the above on (1.0 in. × 1.0 in. × 0.064 in.). The diameter of the nozzle which is located at the center of the cold plate is 0.022 in. Again, a round heater (0.15 in.^{2}) as one given in Figure 9.10 was installed at the center of the cold plate.

The cold plate is tested at the vertical orientation as shown in Figure 9.12. The water inlet is at 24°C. The measured cold plate temperatures on the heater side

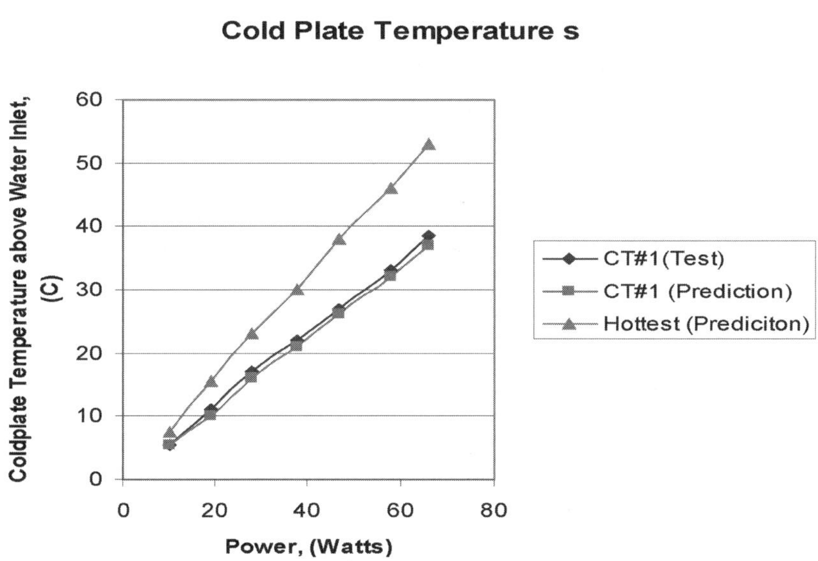

FIGURE 9.11 Temperatures of microchannel cold plate.

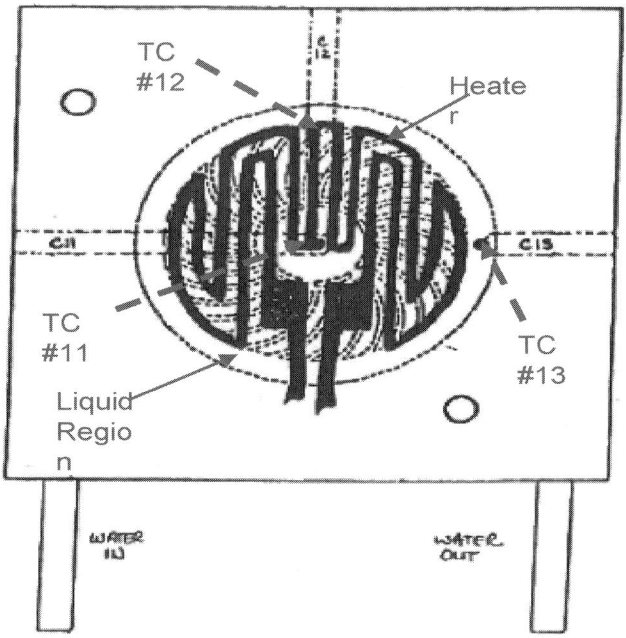

FIGURE 9.12 Swirl cold plate.

at various water flow rates ranging from 33 ml/min to 73 ml/min are presented in Figure 9.13. The open symbols (TC 11) represent the temperature at the center of the cold plate. The hottest spot (TC12) represented by a closed symbol is located near the center of upper half of the cold plate. For a flow rate of 33 ml/min, the measured hottest spot is about 40°C above the water inlet temperature when 50 W of heat are applied. This result is similar to that of the brass cold plate discussed previously. However, the pressure drop in the swirl cold plate is much lower than that of the brass cold plate. The heat transfer coefficient and the pressure drop are correlated into the following equations, respectively.
For center location

$$Nu_d = 0.171\, Re_d^{0.6} \tag{9.6a}$$

For hottest spot location

$$Nu_d = 0.155\, Re_d^{0.6} \tag{9.6b}$$

For pressure drop

$$\Delta P = -0.034 + 0.2417 \times 10^{-3}\, Re_d + 0.1299 \times 10^{-5}\, Re_d^2 \tag{9.7}$$

where Re_d is Reynolds number based on the nozzle diameter.

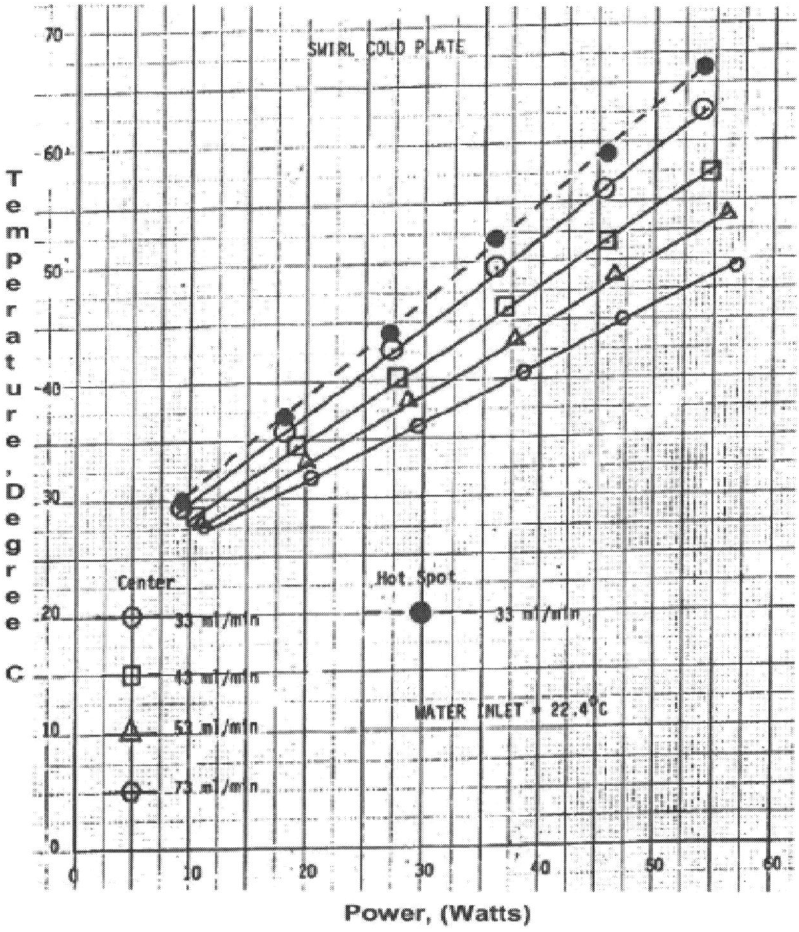

FIGURE 9.13 Thermal performance of swirl cold.

The microchannel coolers proposed by Tuckerman [12, 13] could not be fabricated easily and also pressure drops were extremely high which were not suitable for the practical application in the real world conditions. Recently, the advances in manufacturing technologies, including the progress in high-rate, deep reactive ion etching of silicon have made possible for the easy fabrication of silicon microchannel cooler at the reasonable cost [15]. In addition, the methods for reducing the pressure drop have also been reported, including subdividing the flow into multiple heat exchanger zones with shorter channel lengths and manifold designs with large cross sectional area is presented in Figure 9.14. For flow rate at 1.01 l/min, the overall thermal resistance and the pressure drop are 17.5 °C mm^2 /W and 30 kPa, respectively.

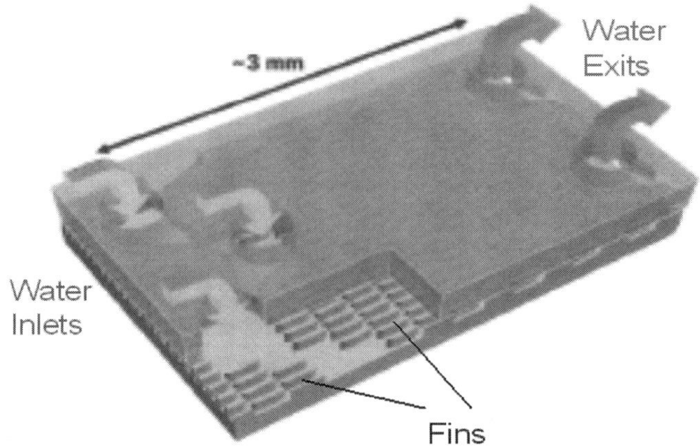

FIGURE 9.14 IBM microchannel cooled silicon chip.

Yeh [16] conducted an experiment to study the heat transfer and flow friction of offset fins as shown in Figure 9.15 with various liquids in narrow passage. The fin stock used in cold plate slat for all heat transfer tests has 15 fins per inch. All fins have a height of 0.04 in. (channel height), a thickness of 0.004 in., and a length of 1.25 in. in the flow direction.

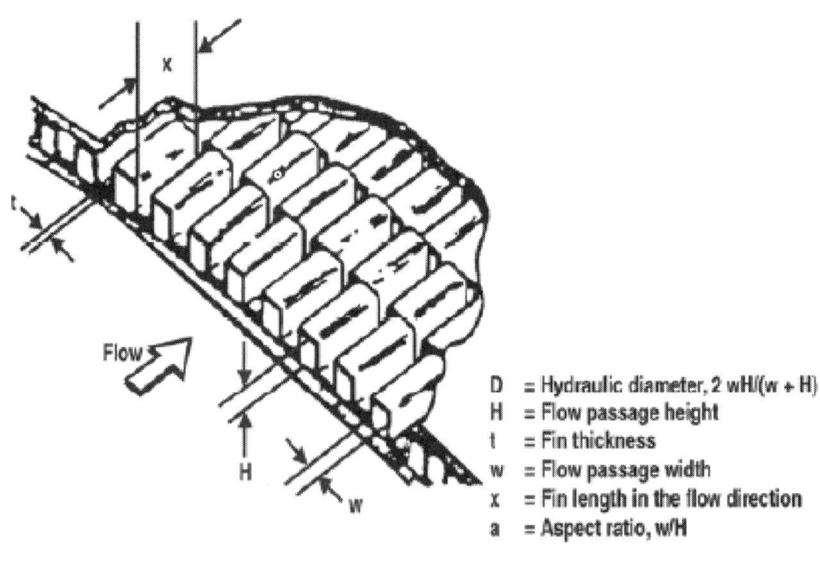

FIGURE 9.15 Thin rectangular offset fins.

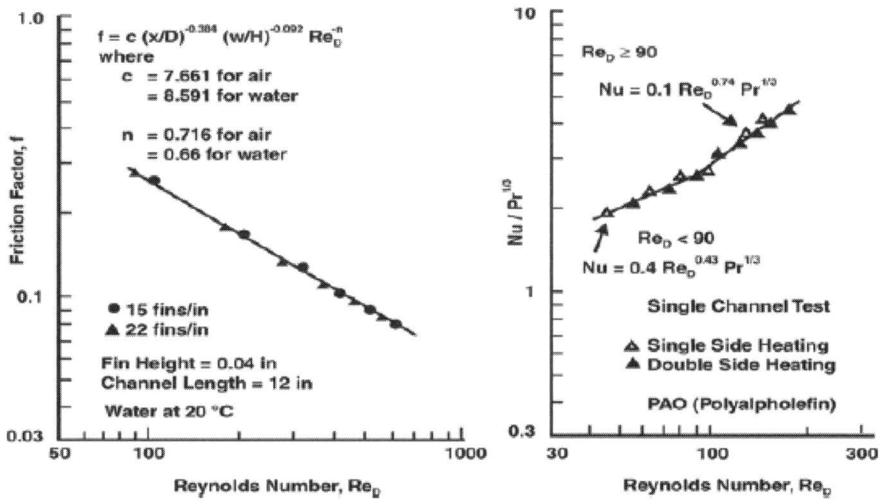

FIGURE 9.16 Friction factor and Nusselt number for offset fins in narrow passage.

The summary results are presented in Figure 9.16 and are correlated into the following equations:

$$f = C(x/\mathrm{D})^{-0.384}(w/H)^{-0.092}Re_{\mathrm{D}}^{-n} \tag{9.8}$$

and

$$Nu = 0.1\ Re_{\mathrm{D}}^{0.74}\ Pr^{1/3} \text{ for } Re_{\mathrm{D}} \geq 90 \tag{9.9a}$$

$$Nu = 0.4\ Re_{\mathrm{D}}^{0.43}\ Pr^{1/3} \text{ for } Re_{\mathrm{D}} < 90 \tag{9.9b}$$

where C and n are 8.591 and 0.66 for water, respectively, and

$Nu(h\mathrm{D}/k)$ = Nusselt number
$Re_{\mathrm{D}}(\rho V\mathrm{D}/\mu)$ = Reynolds number based on hydraulic diameter D
$Pr(\mu C_p/k)$ = Prandtl number

Bland, Niggemann, and Parekh [17] developed a compact high intensity cooler (CHIC) to cool 50 W/cm^2 device for space applications. Therefore, the coolant must not freeze in the system and should also have a low pressure drop to reduce the pumping power. In addition, the cooler must also be lightweight and compact. The geometry (cross-section) of this heat exchanger cannot be larger than 1 cm × 1 cm square area. The primary coolant in the system is Freon 11.

The cooler consists of a stack of thin copper orifice plates and spacers which are bonded together and arranged to provide liquid jet impingement heat transfer on successive plates as shown in Figure 9.17. There are cutouts on the spacers in the region of the orifices. The hole pattern on alternate orifice plates is offset by half the hole pitch so that jets from orifices are targeted at locations midway

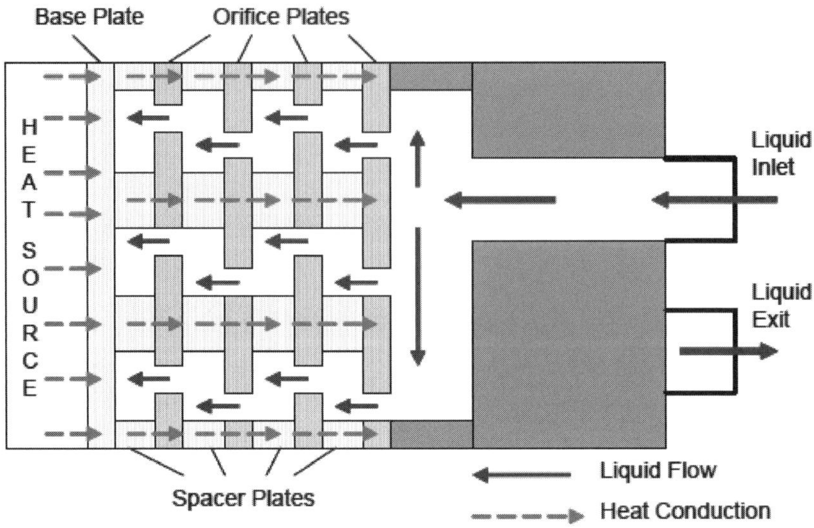

FIGURE 9.17 CHIC heat exchanger design concept.

between holes on the next orifice plate. This stack is bonded to the base plate which directs the liquid to the drain channels and provides the means by which the heat exchanger is attached to the heat source or device to be cooled.

Figure 9.17 shows the fluid entering through an inlet tube into an inlet plenum where it separates into three streams and passes through orifices in the top plate. The liquid jets impinge on the next orifice plate to remove heat conducted through the spacer plates from the heat sources. Heat is first conducted to the base plate of the cooler and then is conducted through contacting areas between the orifice plates and the space plates and is finally transferred to the coolant from the orifice plates. Heat transfer path is shown by the dashed-line arrows, while the flow path is represented by the solid arrows.

The liquid from impinging jets then flows transversely into the nearby orifices and impinges on the next orifice plate again. The process is continued until the jets impinge on the base plate of the drain channels. The coolant then flows laterally through the narrow gap inside the outer shell into the drainage slot and out through the fluid outlet tube.

Thermal tests on the wafer thin coolers [18] were performed for the heat flux ranging from 5 to 125 W/cm^2. The primary application of these liquid cooled devices is to remove heat from compact gallium arsenide (GaAs) diode wafers used in laser communications as shown in Figure 9.18. The tested wafer thin coolers are a double pass microchannel cooler, two types of single pass microchannel coolers, and two versions of jet impingement coolers such as CHIC mentioned above. In a typical application, GaAs diode wafers are fabricated including 40 to 400 individual diode emitters. For maximum efficiency, all diodes must operate below or near 25°C with the maximum temperature gradient less than 1°C.

The coolers are designed to support high heat fluxes over 100 W/cm^2 with minimal temperature gradients at the heating surface. The thin coolers have a heat

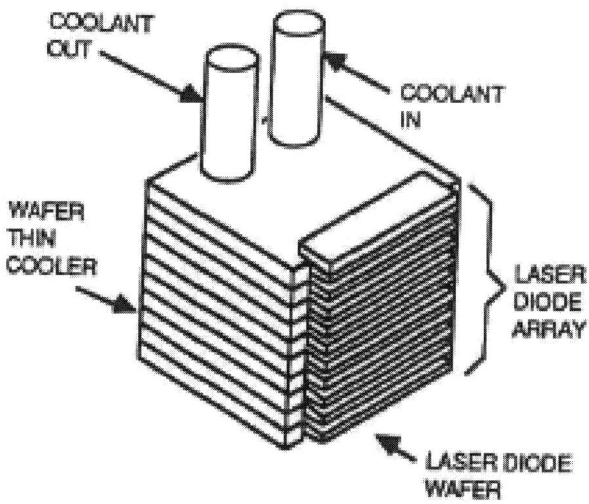

FIGURE 9.18 Laser diode/cooler stack.

input area of 1 cm^2 and the thickness ranging from 1 to 1.8 mm with water as coolant. One single pass cold plate was made of beryllium oxide and other four coolers were made of copper. Typical results for heat flux at 125 W/cm^2 with coolant flow rate of 10 kg/hr indicated that the heat transfer coefficients and maximum surface temperature gradient were 8 W/cm^2 °C and 0.7°C for double pass microchannel, 6.5 W/cm^2 °C and 6.8°C for the copper single pass microchannel, 3.9 W/cm^2 °C and 6.1 °C for the beryllium oxide single pass microchannel, 3.5 W/cm^2 °C and 2.5°C for the first impingement cooler and 5.3 W/cm^2 °C and 2.4°C for the second impingement cooler, respectively.

9.4 INDUSTRY TRENDS IN LIQUID COOLED ELECTRONIC EQUIPMENT

Since most of telecommunication systems with exception of some outdoor equipment are cooled by air, it is of interest in examining the current industry trends in liquid cooled electronic equipment, especially in computer industry. In addition to direct immersion cooling, two general schemes in liquid cooling of chip package are direct internal cooling of silicon die and external cooling of the package through the cold plate. The former includes the microchannel cooling as described by Tuckerman and Pease [12], Tuckerman [13], and Colgan [15] where the cooling channels are integral part of the silicon die. On the other hand, the latter is to replace the conventional heat sinks used in the air cooled systems by a liquid cold plate which is in contact with the component case (or lid). The flow pattern in both the internal and external cooling methods includes liquid through channels and jet impingement.

Water has long been used in cooling of computers. One of most noticed liquid cooled computers in the earlier dates was IBM Thermal Conduction Module (TCM)

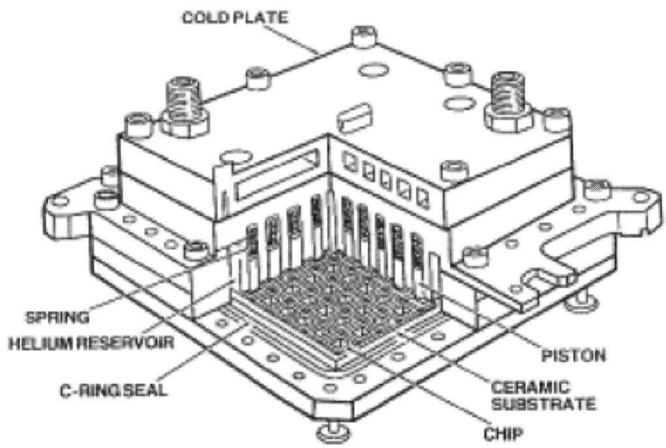

FIGURE 9.19 IBM liquid cooled thermal conduction module.

as shown in Figure 9.19 [19]. IBM TCM is an indirect liquid cooling scheme. Heat generated from microelectronics (chips) is conducted through spring loaded pistons to the cold plates on the top. To reduce the thermal contact resistance, air is replaced by helium to fill up the space in the individual piston because helium has a higher value of thermal conductivity. The construction of TCM is extremely complicated.

To modify the thermal path between the chip/package and the water cooled cold plate, Andros and Shay [20] proposed utilizing a miniature flexible heat pipe to connect the chips and the cold plate as illustrated in Figure 9.20. Similar modification as shown in Figure 9.21 was also made by Clark and Metreaud [21]. The latter also

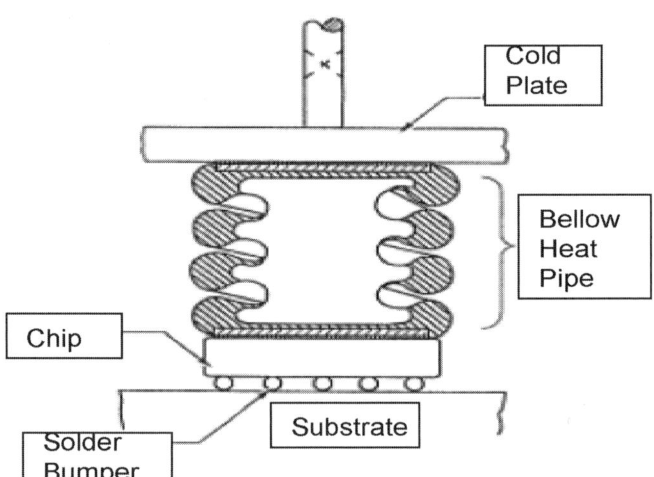

FIGURE 9.20 Liquid cooling of bare chip with flexible heat pipe.

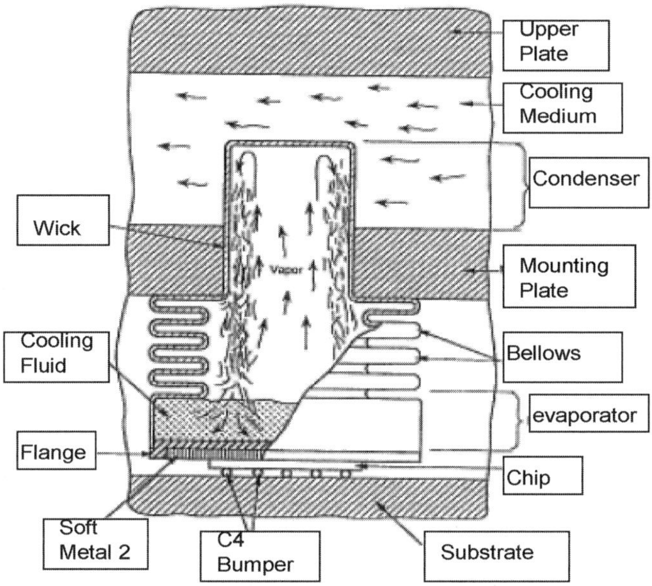

FIGURE 9.21 Liquid cooling of bare chip with flexible heat pipe extension into liquid flow.

eliminates the contact thermal resistance at the interface between the cold plate and the flexible heat pipe by extending the flexible heat pipe directly into the liquid flow.

Liquid jet impingement was employed for cooling a large scale computer (Fujitsu M-780) as given in Figure 9.22 [22]. As shown in Figure 9.22, R_{cond} is the thermal resistance due to conduction from the chip junction to FTC (flexible thermal conductor) heat transfer plate surface, and R_{conv} is the thermal resistance

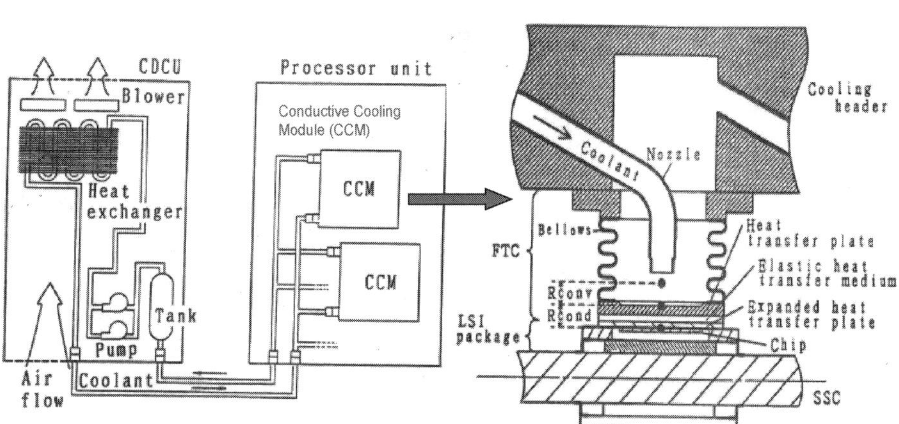

FIGURE 9.22 Fujitsu FACOM M-780 computer.

due to convection (water jet impingement) from the FTC heat transfer plate to the coolant. The total thermal resistance is estimated to be 2.4°C/W with chip power of 9.5 W (336 chips on board). The coolant is supplied by the coolant distribution and control unit (CDUC) as shown in the left side of the figure.

The NEC supercomputer SX-2 as illustrated in Figure 9.23 [23] employs a high-density multi-chip package and a liquid (water) cooling module (LCM) structure which consists of a heat transfer block (HTB), a cold plate and 36 studs placed in machined holes in the HTB to insure good contacts with LSI chip carrier mounted on a multi-layer substrate. The cooling unit as shown on the left of the figure supplies the cooling water to the cold plate of LCM. The average thermal resistance is measured at 3.74°C/W with the maximum system power of 3000 W.

Yeh [11] employed dielectric liquid (coolanol-25) to cool RF modules with a large cold plate as presented in Figure 9.24. Heat generated from RF modules was transferred by heat pipes to the aluminum cold plate consisting of 53 extruded channels with fluid flowing in opposite direction in alternating channels. Each heat pipe serves a pair of RF modules. The total system power is 7.4 kW. The coolant is supplied to the cold plate at seven gallons per minute (GPM) with pressure of 70 PSIG and a maximum temperature of 46°C.

Once coolant enters supply manifolds, the fluid flows in opposite directions through the adjacent fluid channels and is collected in the discharge manifolds as it leaves the system. The counter flow scheme is selected to reduce the maximum cold plate temperature and also to minimize the temperature difference of over the cold plate. Each fluid channel which is made by the extrusion process includes several subchannels in order to enhance the liquid convection heat transfer of the cold plate by increasing the heat transfer surface. As presented in Figure 9.24,

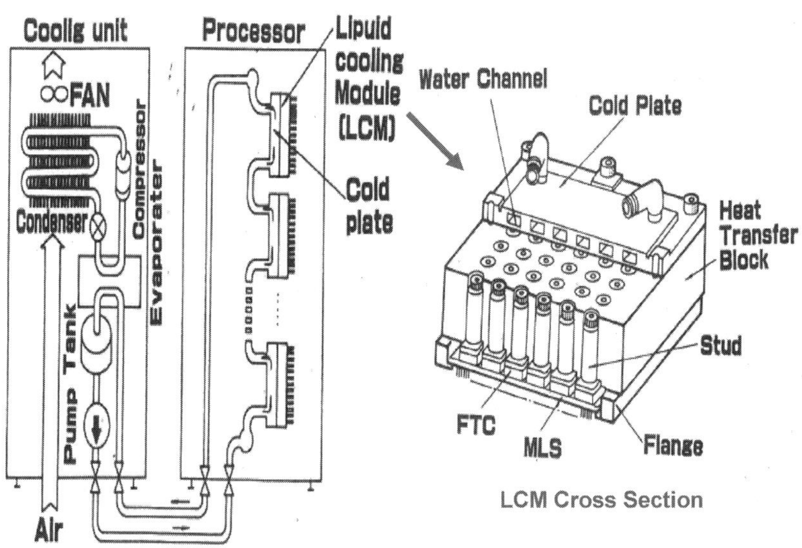

FIGURE 9.23 NEC SX-2 cooling system.

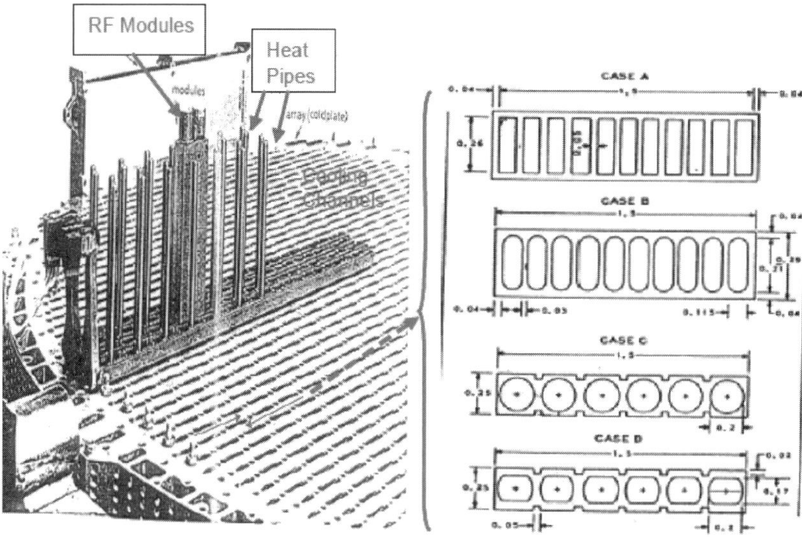

Section of Cold Plate with 53 Counter Flow Channels

Subchannels Inside A Single Channel

FIGURE 9.24 Liquid cooled clod plate for RF systems.

FIGURE 9.25 IBM Power 575 system.

various subchannels in each case (from top, Case A to bottom, Case D) are investigated to meet the following requirements : (1) thermal performance, (2) weight, (3) producibility, (4) structural integrity and (5) cost. The initial design of 10 rectangular subchannels (Case A) is not acceptable because the sharp, square subchannels are not feasible in the extrusion process and Cases B and C are rejected because the wall thickness is too thin.

The hybrid cooling over the boards has recently received considerable attention. In other words, the high power components on the board are cooled by the liquid while the rest components are still cooled by the air. Figures 9.25 [24] and 9.26 [25] present some examples of this cooling scheme. The former corresponds to IBM Power 575 system while the latter is Fujitsu GS 8900. The issue of possible condensation inside the rack must carefully be considered for the system with hybrid cooling.

The possible fan failure and also the acoustic noise issue may still exist in the hybrid cooling scheme. In addition, the liquid loop (tubes) may become obstacle to the certain air flow to the components. Therefore, the active liquid cooling is the best approach to achieve the energy efficiency both at equipment level and data center and also to support the much higher system power while eliminating the acoustic noise from air cooling.

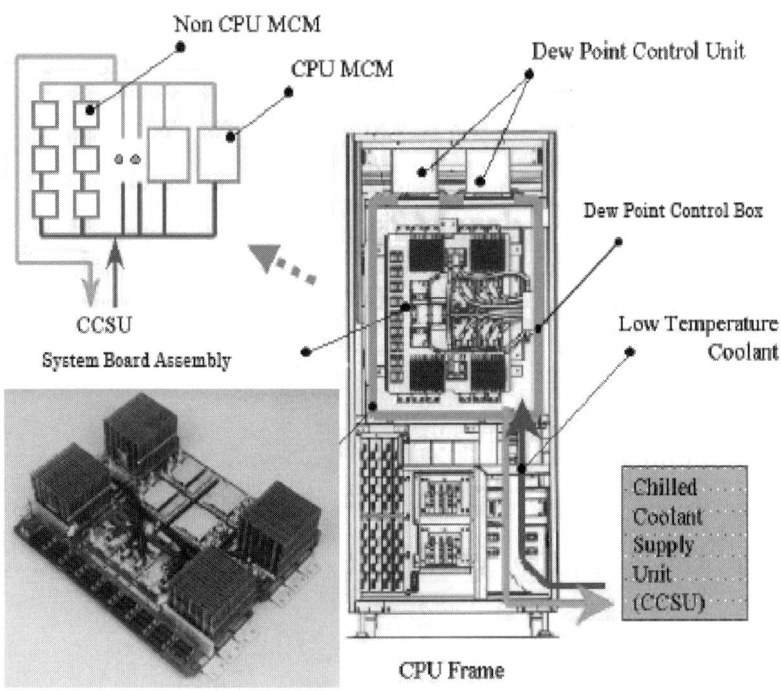

FIGURE 9.26 Fujitsu High End Server (GS8900).

Figure 9.27 IBM Zurich liquid cooled board.

IBM Research Zurich Lab in conjunction with ETH developed a warm water (60°C) cooled supercomputer (Aquasar) using microchannel cold plate for cooling the components as shown in Figure 9.27 [26]. The detailed cooling scheme of individual cold plate developed by Escher [27, 28] is given in Figure 9.28. For a flow rate of 1.3 l/min, the total thermal resistance between the maximum heater temperature and fluid inlet temperature is 0.09 cm^2 °C/W with a pressure drop of 0.22 bar on a 2 × 2 cm^2 chip.

Careful planning of the fluid network must be made to insure the proper flow rate to individual boards and also to individual components on given boards to avoid overheating. The former is related to the system flow distribution and the later represents the flow distribution on the board level.

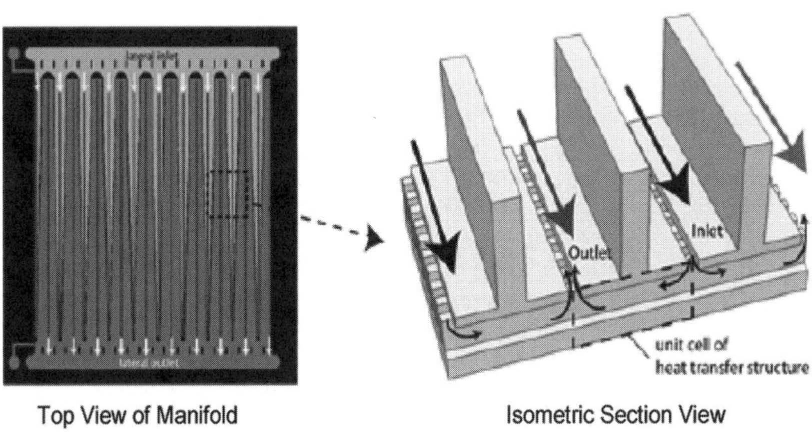

Figure 9.28 Scheme of a microchannel heat sink.

9.5 SYSTEM THERMAL DESIGN

While the major effort during the design of any liquid cooled electronic systems are at the cooling aspects, attention must also be given to the practical design aspects which are reflected in the maintainability of the equipment. To achieve ease of maintenance and repair, system designs should be simple, and provide easy access to the internal parts. Detailed maintainability studies are required for any equipment based on selected cooling concepts. Factors requiring consideration include: (1) a minimum of external connections for the cooling circuitry, (2) quick dis-connect devices, (3) enclosures which are easy to disassemble and/or remove from the working electronic equipment, (4) mounting devices from which the electrical components are easily removed, and which are easily installed and removed from the chassis, and (5) seals for use with pressurized, immersion and vaporization cooling concepts. In addition, the system should not be required to avoid drain the coolant during the service.

The degree of maintainability achieved of given equipment depends not only on the simplicity of its design and accessibility of its components but also on the reliability of the system/components. Therefore, the advanced cooling concepts are more complex and less accessible than those using air cooled systems, this does not necessarily imply that such equipment requires more maintenance. The increase in reliability expected from the liquid cooling should reduce the frequency of required repairs which result in the greater availability of the equipment for use despite the fact that it may take a longer time for a repair as compared with the air cooled systems.

9.5.1 Direct Liquid Cooling

As in air cooling, liquid cooling can be either direct or indirect cooling. Dielectric liquids must be utilized in the direct cooling because liquids are in a direct contact with electronic components. Direct cooling can employ natural convection/pool boiling or forced convection/flow boiling mode of heat transfer with either single- or two-phase flow as shown in Figure 9.29A and 9.29B [29]. The electronic components are immersed in a pool of liquid in the former case and the liquid is forcefully circulated over electronic components in Figure 9.29B.

In the simplest case as shown in Case A of Figure 9.29A, an electronic equipment is immersed in a sealed contained. Heat generated by electronics is removed by liquid through natural convection and is further carried away to contact with the walls of the container and is finally cooled by transferring heat to outside of the container. The flow circulation is caused by the density difference between the cold and hot fluids. This process is often referred to as a natural circulation. In addition, the heat sink as given in Cases B and C is actually a condenser to condense the vapor back to liquid. It should be noted for liquid filled containers as given in Figure 9.29A, account must be taken of the volumetric change due to its temperature rise.

If forced convection immersion cooling is used, the heat removed from the equipment is rejected to the secondary fluid by means of a suitable heat exchanger. The higher heat transfer coefficient associated with the forced convection significantly improves the system heat removal capacity. However, such forced convection

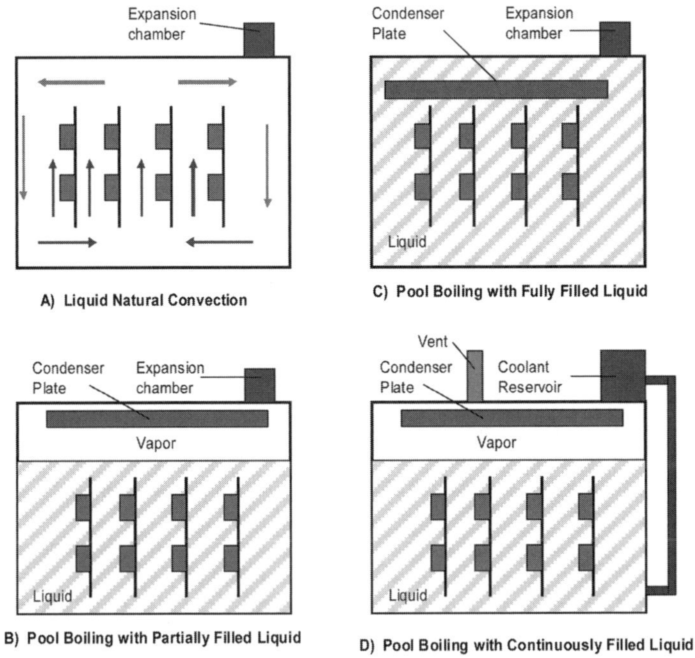

FIGURE 9.29A Passive direct liquid cooling.

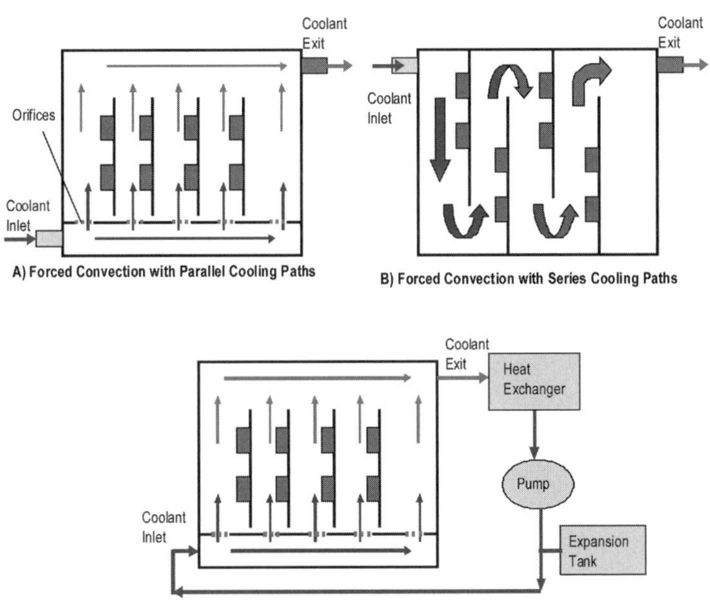

FIGURE 9.29B Active direct liquid cooling.

system requires an additional pump and heat exchanger which increases the overall power consumption, weight and cost. An accumulator is often added to the system to allow for liquid expansion and minimize vapor lock.

For forced convection and/or flow boiling as presented in Figure 9.29B, the flow pattern is parallel in Case A while is in series in Case B. The parallel flow arrangement is preferred and it has many advantages over the case with a series flow scheme. The major advantage of the parallel flow system is to provide minimum pressure drop and maximum distribution of the cold liquid. The system pressure drop is extremely small for the parallel flow arrangement as compared to the series flow system because of small flow rate (fraction of the total flow) and short flow length of each parallel flow path. In addition, the temperature difference among all boards is relatively small. On the other hand, the fluid is continuously heated in the series flow arrangement which results in a large the temperature difference between the inlet and exit. It should be noted that the pressure drop through each of the parallel path is essentially the same. In order to provide sufficient coolant to each flow path, the flow rate to each path (boards) must be proportional to the heat load of individual paths (boards). Therefore, the flow rate to individual paths must be properly regulated.

The nucleate boiling will eventually take place in either natural or forced convection liquid immersion if the system power is sufficiently high. The nucleate boiling is the most effective heat transfer mechanism. In addition to extremely high heat transfer coefficients in this region, the boiling conditions are characterized by relatively small temperature differences between the heating surfaces and the coolant, typically less than 5°C. Therefore, one should take advantage of this efficient heat transfer mechanism in the design of liquid cooling systems for electronic equipment. One major consideration in selection of the coolant for this type system is the boiling point of the fluid. The boiling temperature of a fluid is a function of the saturated pressure of the system which must be lower than the required temperature limits of electronic components used in the equipment. The desirable boiling point of given fluids for the system under consideration can be estimated by the following equation

$$T_{bp} = T_j - Q^* \times \theta_{j-c} - \Delta T_{sat} \tag{9.10}$$

where

T_{bp} = liquid boiling point
T_j = component junction temperature limit
Q^* = component power
θ_{j-c} = component junction-to-case thermal resistance
ΔT_{sa} = difference between component case and liquid saturated temperatures

9.5.2 Indirect Liquid Cooling

In an indirect cooling, the coolant does not make a direct contact with the electronic components. Heat is removed from electronics by various heat transfer modes to a cooling device such as cold plate, and heat is further carried away to another heat exchanger where the hot liquid is cooled and is returned to the cold plate. The liquid cooling process is repeated again and again. A condenser will be required if the boiling takes place in the cold plate. The thermal resistance at the interface between the cold plate and components is extremely critical in any indirect liquid cooling system and care must be made to minimize this interface thermal resistance. Indirect liquid cooling has certain advantages over immersion or direct liquid cooling. Among them are easier accessibility for maintenance, less possibility of fouling equipment, less concerns of handling of coolants, and great flexibility on coolant selections where, for example, dielectric fluids are no longer required and high heat transfer fluids such as water can be used.

A single fluid heat exchanger designed for removing heat from any heat sources is called a "cold plate." The cold plate is a special case of a heat exchanger which involves two fluids in the system. The cold plate can be employed in either air or liquid cooling systems. Cold plate cooling concepts take advantage of the large number of efficient heat exchanger configurations, e.g., various types of fins. As in any heat transfer systems, the key factor in thermal design of a cold plate is to maximize heat transfer rate and to minimize the pumping power.

9.5.3 General Design Considerations

Though liquid cooling has long been employed in the computer industry for a long time, no liquid cooled indoor telecommunication equipment is commercially available at the present time. Water is the most common fluid used in the high power computers because of its high heat transfer capability. However, water can not be used in a direct immersion cooling and nor for outdoor applications. The former is due to its poor dielectric properties and the later is because of its high freezing point. Water also must be properly treated prior to being used in any cooling systems. Most chemicals used in water treatment are proprietary formulas. It is very important to select a compatible pair of liquid coolant and wetted materials throughout the entire liquid loop. For material compatibility, copper must be used for the entire liquid loop if water is adopted as the coolant. On the other hand, aluminum can be employed for all dielectric fluids, including Glycol. De-ionized water which is often used as coolant is compatible with stainless steel. All above three types of materials (copper, aluminum, and stainless steel) are compatible with Glycois.

All cooling methods discussed previously only deal with liquid cooling of individual components or boards (subsystems). Generally, coolant first goes into an inlet or supply manifold once it enters the system (equipment) and then flows through inlet connecting tubes to the cold plates associated with individual boards and finally liquid is collected at the exit or discharge manifold through exit connecting tubes prior to leaving the system. Since the system (equipment) always consists of a number of boards (or subsystems) whose power may vary, one must make sure proper flow rate be distributed to individual boards. The flow rate to individual boards should be properly regulated in proportion to the power of the boards. Sometimes, a large flow rate will still be needed even for a low power board if this board includes some critical components such as optical modules

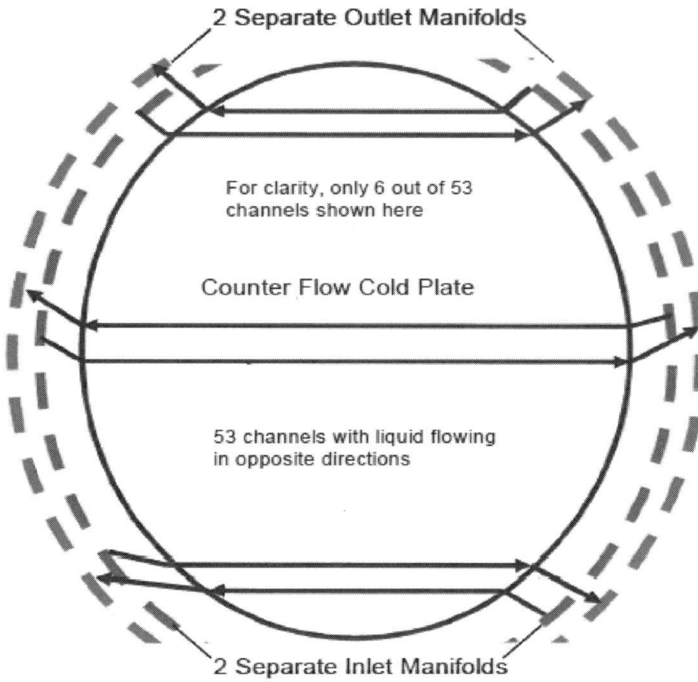

Figure 9.30 Redundant liquid cooling system.

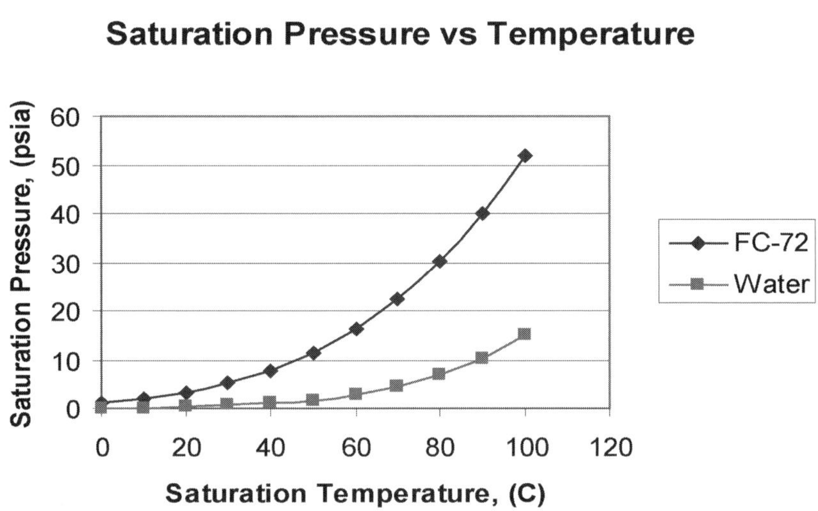

Figure 9.31 Saturation pressure versus temperatures for water and FC-72.

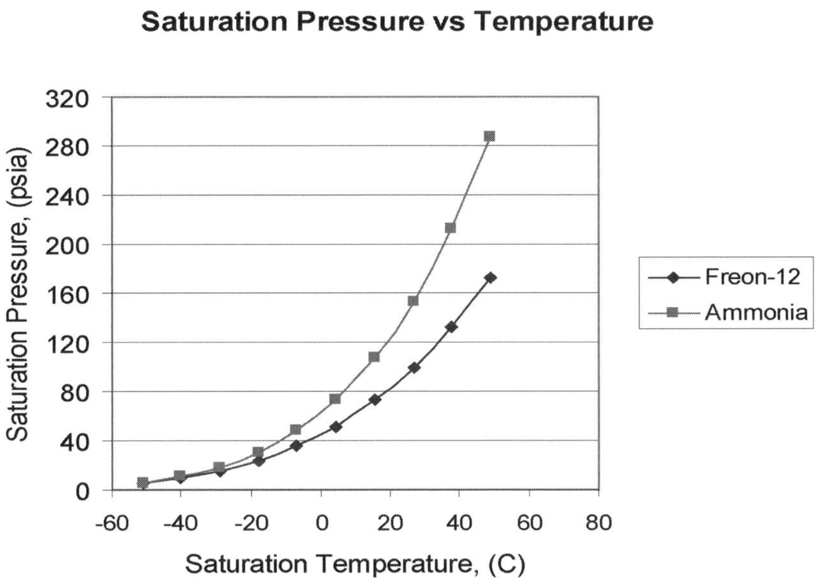

FIGURE 9.32 Saturation pressure versus temperatures for Freon-12 and ammonia.

with low component temperature limits. The simplest way to have a proper flow distribution to individual boards is to place various sizes of the orifices in the inlet connecting tubes or exit connecting tubes.

In addition, the equipment must also be designed with redundant cooling loops to insure the system reliability though it makes system more complicated. One of examples of the redundant cooling loops is presented by Yeh [11]. The sketch of the redundant cooling system for the cold plate as shown in Figure 9.24 is given in Figure 9.30. The system has two independent and separate flow paths through the entire cold plate.

The above discussion of indirect liquid cooling is limited to the single-phase liquid cooling. Two-phase flow cooling has much higher thermal efficiency than that of single-phase liquid cooling. However, the two-phase flow system is much more complicated which requires a condenser so that the vapor can be condensed for reuse. For water cooled systems, including all current supercomputers, water must be boiled at around 60°C in order to maintain the chip junction temperature below 85°C. The boiling point of water at atmospheric pressure (14.696 psia) is 100°C. Therefore, system pressure must be adjusted to below the atmospheric pressure at about 2.9 psia as given in Figure 9.31 by pulling the vacuum. Figure 9.31 shows the saturation pressure as a function of the saturation temperature for the water and the FC-72. As can be seen from the figure, the boiling point for FC-72 at the atmospheric pressure is at about 61°C which is much lower than that of water (100°C). Similar curve for refrigeration fluids such as Freon-12 and Ammonia is given in Figure 9.32.

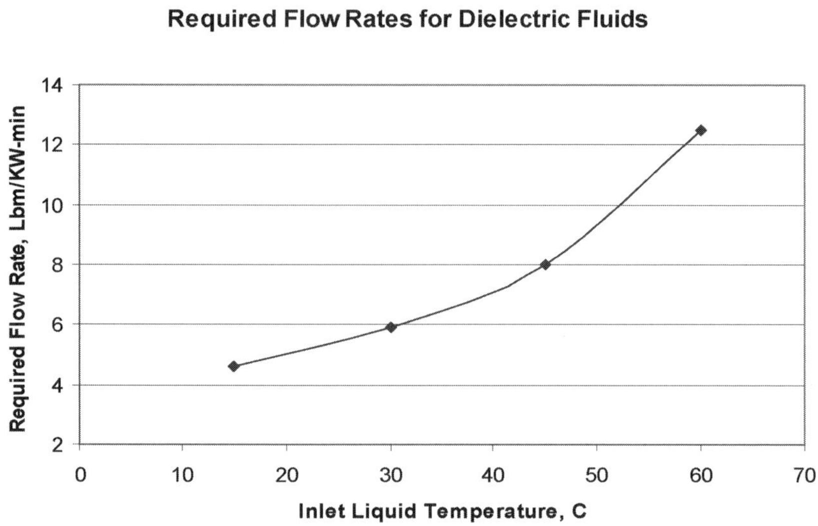

FIGURE 9.33 Estimated flow rates required of dielectric fluids for given system power.

Generally, the effects of coolant properties on heat transfer and friction factor is minimum because of the temperature difference between the inlet and exit of a cold plate is not large enough to be concerned. However, it may become critical in the microchannel cooling systems. This is due to the fact that very little flow is used in such systems which result in large temperature gradient of the fluid between the inlet and the exit. The microchannel liquid cooling technologies are generally employed in cooling of high power systems and it typically has a large pressure drop because of extremely small channel cross section even operating at small flow rates. With these factors, high system power and high pressure drop, the liquid coolant could be boiled near or at the exit of the cooling systems.

For any forced convection system, the pressure drop is as important as the heat transfer. A high-pressure drop system will require a large pump which not only consumes more power but also is heavy and costly. High heat transfer always accompanies by the high pressure drop. A good system design must have a good balance between the heat transfer and the pressure drop. Therefore, the goal of system thermal design is to maximize heat transfer and to minimize pressure drop. The best way to achieve the goal is to adopt efficient cooling schemes which will require the least flow rate. For dielectric fluid systems, the required system flow is proposed and is given in Figure 9.33. This curve is generated based on the limited data by Yeh [11] and the past experience of the authors. The actual flow rate required is totally depending on the system thermal requirements and the selective cooling methods. Generally, the higher flow rates are needed for higher liquid inlet temperatures. The proposed curve is intended to be used as a general guideline in determining estimated flow rates required for the given system power. The much less flow rates will be needed if the coolant is water because of its high heat transfer capacity. To limit power consumption of the pump, it is recommended that the system pressure drop should exceed 20 psia under the supplied flow rate and the coolant inlet temperature.

REFERENCES

1. Yeh, L. T. and Chu, R. C., "Thermal Management of Microelectronic Equipment", ASME Press, 2002.
2. Knight, A. F., "Choice of Fluids for Cooling Electronic Equipment", Electro-Technology, June, 1963.
3. Kelly, D. J., "Resistance of Materials to Hydraulic Fluids", Machine Design, January, 21, 1972.
4. Lee, T. Y., Saylor, J., Simon, T., Tong, W., Wu, P. S., and Bar-Cohen, A., "Fluid Selection and Property Effects in Single- and Two-phase Immersion Cooling", Proceedings of Intersociety Conference on Thermal Phenomena in Electronic Components, May, 1988.
5. Park, K. A., and Bergles, A. E., "Natural Convection Heat Transfer Characteristics of Simulated Chips", Heat Transfer in Electronic Equipment, ASME HTD-48, 1985.
6. Marto, P. J. and Lepere, V. J., "Pool boiling Heat Transfer from Enhanced Surfaces to Dielectric Fluids", J. of Heat Transfer, Vol. 104, 1982.
7. Oktay, S., "Departure from Natural Convection (DNC) in Low-Temperature Boiling Heat Transfer Encountered in Cooling Micro-Electronic LSI Devices", Proceedings of 7th International Heat Transfer Conference, 1982.
8. Yeh, L. T., "Analytical Solutions for a Counterflow Heat Exchanger with Space Dependent Wall Heat Dissipation," Heat Transfer in Electronics Equipment, ASME HTD-48, 1985.
9. Yeh, L. T., and Gingrich, W. K., "Numerical Solutions for a Multiple-Channel Counterflow Heat Exchanger with Space Dependent Wall Heat Dissipations", Proceedings of the 8th Int. Heat Transfer Conference, 6, San Francisco, CA, 1986.
10. Gingrich, W. K., Yeh, L. T., and Fuhr, T. D., "Transient Temperature Distribution for a Multiple-Channel Counterflow Heat Exchanger with Space Dependent Wall Heat Dissipations", ASME Winter Annual Meeting, San Francisco, CA, 1989.
11. Yeh, L. T., "Thermal Design of a Multiple-Channel Bidirectional-Flow Cold Plate for Solid-State Phased Array Radars", Proceeding International Symposium on Cooling Technology for Electronic Equipment, Honolulu, HI, March, 1987.
12. Tuckerman, D. B. and Pease, F. F., "High Performance Heat Sinking for VLSI", IEEE Electron Devices Letter, EDL-2, 1981.
13. Tuckerman, D. B., "Heat Transfer Microstructures for Integrated Circuits", Doctoral Thesis, Stanford University, 1984.
14. Yeh, L. T., Unpublished data, 1989.
15. Colgan, E. V., Bezama, R. J., Gaynes, M., and Marston, K. C., "A Practical Implementation of Silicon Microchannel Coolers", Electronics Cooling, Volume 13, Number 4, 2007.

16. Yeh, L. T., "An Experimental Study of Offset Fins in a Narrow Channel", Proceedings of 1st International Conference Aerospace Heat Exchanger Technology, Palo Alto, CA, 1993.
17. Bland, T. J., Niggemann, R. E., and Parekh, M. B., "A Compact High Intensity Cooler (CHIC)", 13th Intersociety Conference on Environmental System, July 11–13, 1983.
18. Grote, M. G., Hendron, R. E., Kipp, H. W., Lapinski, J. R., "Test Results of Wafer Thin Coolers at Heat Fluxes from 5 to 125 W/cm^2 °C", 18th Intersociety Conference on Environmental Systems, San Francisco, CA, July 11–13, 1988.
19. Oktay, S. and Kammerer, H. C., "A Conduction Cooled Module for High Performance LSI Devices", IBM J Research Development, 26 (1), 1982.
20. Andros, F. E., and Shay, R. J. E., "Micro Helix Thermo Capsule", US patent No. 4,313,492, February, 1982.
21. Clard, B. T. and Metreaud, C. G., "Cooling device for Multilayer Ceramic Modules", IBM Technical Disclosure, 20 (5), 1977.
22. Yamamoto, H., Udagawa, Y., Suzuki, M., "Cooling System for FACOM M-780 Large-Scale Computer", Proceeding International Symposium on Cooling Technology for Electronic Equipment, Honolulu, HI, March, 1987.
23. Mizuno, T., Okano, M., Matsuo, Y., and Watari, T., "Cooling Technology for The NEC SX Supercomputer", Proceeding International Symposium on Cooling Technology for Electronic Equipment, Honolulu, HI, March, 1987.
24. Ellsworth, M. J., Jr., Campbell, L. A., Simons, R. E., Iyengar, M. K., Schmidt, R. R., Chu, R. C., "The Evolution of Water Cooling for IBM Large Server Systems : Back to Future", Itherm, 2008.
25. Wei, J., "Hybrid Cooling for Fujitsu Large Computer Systems", 2010 IEEE SPJW, Kyoto, Japan, 2010.
26. Wrunschwiler, T., Meijer, G. I., Paredes, S., Escher, W., and Michel, B., "Direct Waste Heat Utilization from Liquid-Cooled Supercomputer", Proceedings of the 14th International Heat Transfer Conference, Washington, DC, August 8–13, 2010.
27. Escher, W., Michel, B., and Poulikakos, D., "Efficiency of Optimized Bifurcating Tree-like and Parallel Microchannel Networks in the Cooling of Electronics", Intl. J. of Heat and Mass Transf. 52, 2009.
28. Escher, W., Wrunschwiler, T., Michel, B., and Poulikakos, D., "Experimental Investigation of an Ultra-thin Manifold Microchannel Heat Sink for Liquid-Cooled Chips", Intl. J. of Heat and Mass Transf., 2010 (current under review).
29. Helenbrook, R. G., and Anthony, F. M., "Development of Liquid Cooling Techniques for Advanced Airborne Electronic Equipment", Ball Aerospace Company, AFFDL-TR-71-129, March, 1972.

Chapter 10

THERMAL MANAGEMENT AND ENERGY EFFICIENCY

The network traffic in telecommunication industry has grown very rapidly every year since its inception. As projected, the network traffic demand will reach tens or hundreds of Tb/s in a couple of years. In response to the projected growth, new design and architecture are needed in order to face the power-density challenges in the next generation telecommunication networks. As pointed out by Vukovic [1], there are two aspects of these issues. One is the network architecture and another is telecommunication equipment design.

The main goal of the architecture is to minimize the total power per network throughput. To achieve this goal, one can employ the so called all-optical network (AON) which keeps the transfer of bits in the optical domain. The AON can develop a "less power hungry" network with the added benefit of being able to cope successfully with the expected network growth and cost reduction requirements. The second issue deals with the telecommunication equipment design. The key for the successful design is to explore techniques and technologies for reducing power at the source by integrating state-of-the-art industry advances in high-density packaging and high speed interconnects. The key challenge of high-density and high speed equipment is the thermal management which must include a balanced combination of materials, technologies and cost.

The cost of thermal control unit is typically a very small percentage of the total system cost of any electronic equipment. For example, the cooling cost ranges from less than 1% for the basic PCs and to 3–5% for large servers and approaching 10% of largest super computers, especially for the liquid cooled systems. Figure 10.1 shows the cooling cost versus power [1A]. As expected, the higher system power, the more complicated and expansive the thermal solutions will be.

This chapter is to serve as the summary of the book which describe the current industry trends of thermal management in the wide range of telecommunication systems and the data centers.

10.1 POWER DENSITY OF TELECOMMUNICATION EQUIPMENT

Currently, telecommunication equipment is facing numerous challenges. The major one is the increased power density. Architects of the next-generation core

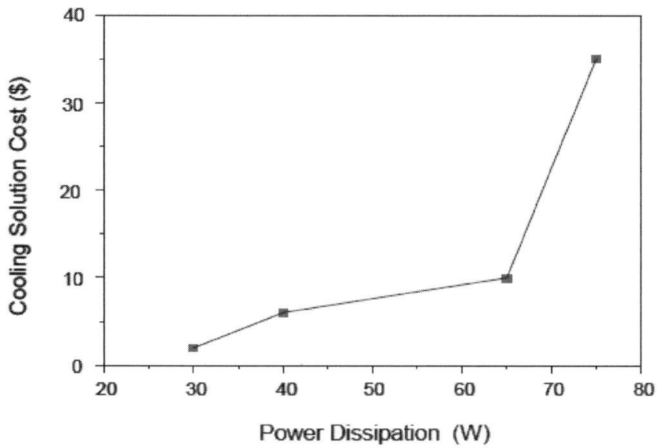

FIGURE 10.1 Cooling cost versus heat dissipation.

network are demanding exponentially higher system throughputs to keep up with network traffic growth. With projected node data transfer rates from 40–100 Gb/sec currently to 200–400 Gb/sec in the next couple of years, and extrapolating from current equipment, future nodes would consume and dissipate up to 100s of kilowatts of power.

Because of the advances in circuit and component technologies, electric circuits have become more efficient and thus heat dissipation from individual transistors has also become less. However, miniaturization of circuits greatly decreases the size of individual devices and increases the number of such devices that can be

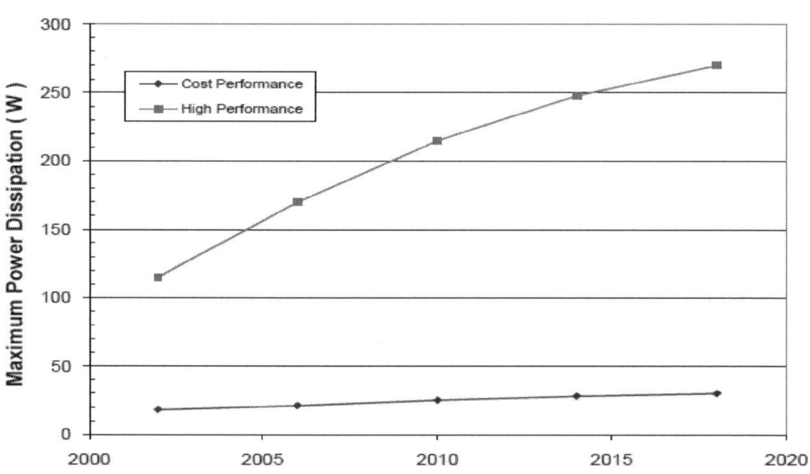

FIGURE 10.2 Maximum chip power trends in telecommunication applications.

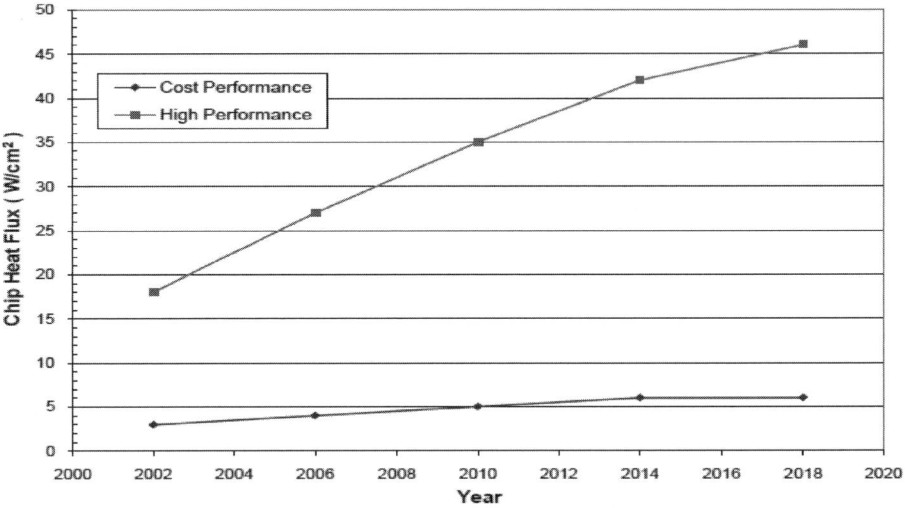

FIGURE 10.3 Maximum chip heat flux in telecommunication applications.

integrated on a single chip. The net result is that the chip heat dissipation has significantly increased as illustrated in Figures 10.2 and 10.3, respectively [2].

The trend reflected in the projected growth in the heat flux based upon the footprint area of the equipment is shown in Figure 10.4 [3]. As can be seen from the figure, the system power for the core network is the highest among all products presented in the figure.

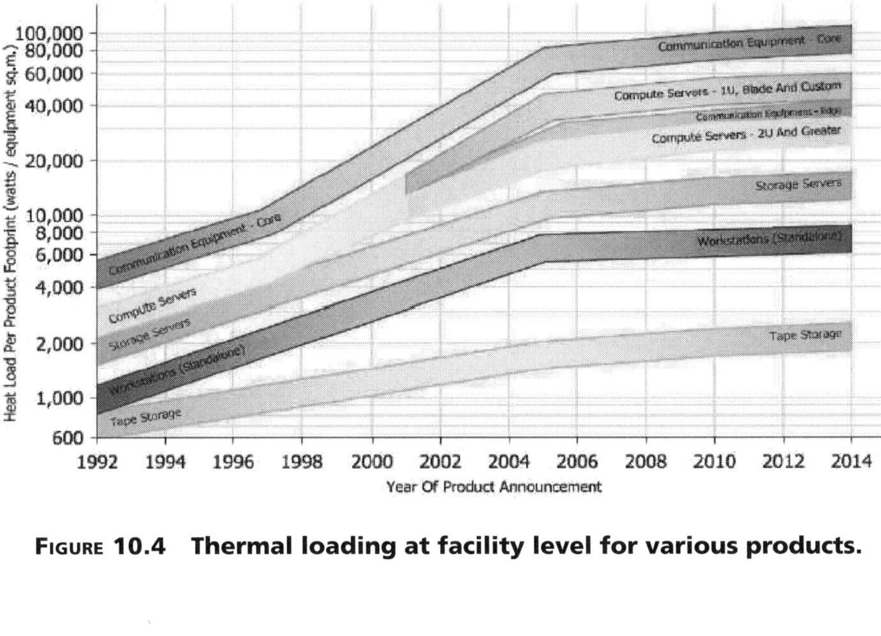

FIGURE 10.4 Thermal loading at facility level for various products.

10.2 CRITICAL THERMAL ISSUES

The equipment power consumption is continuously increased at a fast pace. Since the current telecommunication systems are still focusing at the air cooling, large high performance fans are becoming a must for the high power systems in order to provide the necessary airflow rates. In addition to the system thermal issues, two major concerns about these large fans are the power consumption and the acoustic noise of the fans. The former is related to the equipment energy efficiency, while the latter is strictly regulated by the government and industry standards such as NEBS in USA and North America, and the ETSI in Europe.

The increased system power results in a significant increase in the power consumption and the cost of powering and cooling of the equipment as well as the operation cost of the data centers in the last few years. For example, AT&T and Verizon have reportedly spent over 1 billion and 500 million US dollars, respectively to operate their data centers throughout the United States in 2006. The cost of electricity has steadily increased in the last few years. The operation cost will be even higher due to the expansion of the date centers. In short, the total operation cost could easily be doubled in the next few years. The similar trends are also found in the worldwide base. The operation cost is the major concern for the data centers. Therefore, the energy efficiency becomes the top priority for the equipment manufacturers and also the data center operators.

10.2.1 Equipment Energy Efficiency

The COP (coefficient of performance) is often adopted in the telecommunication industry to measure the thermal or energy efficiency of the equipment. The COP is defined as the ratio of the total system power, including power from electronics, power supply/converters, and fans to the total power consumption of fans. Therefore, the higher power consumption of the fan, the lower system COP will be. In general, the COP of any equipment under the worst-case condition with fans at full speed should be greater 10.

10.2.2 Equipment Acoustic Noise

The equipment for the indoor applications must meet the GR-63 requirements of NBES [4] for the USA and North American markets. For the acoustic noise limits, the GR-63 states that the acoustic noise of any equipment must be less than 78 dBA (sound power) at the ambient of 27°C. On the other hand, ETSI states that the acoustic noise (sound power) should be limited to 72 dBA or less at the ambient of 23°C.

The high acoustic noise is a byproduct of high power systems associated with the fans. The acoustic noise can be reduced by either active or passive solution or combination of both. The passive solution is achieved by adding acoustic noise reduction (absorption) materials at the proper locations throughout the system or installing the acoustic door for the racks. Since the characteristics of the acoustic noise are very similar to that generated by a microphone-speaker, it is possible to create an identical sound wave with 180° phase difference to cancel the fan acoustic noise. Based on this finding, the ANC (active noise cancellation) device which generally includes several electronics on a small printed circuit board, the speaker and the microphone is developed. An ANC device is typically to be installed at the downstream of the fans.

10.3 THERMAL MANAGEMENT AND ROADMAP

Air cooling has long been considered and adapted for cooling of electronics. This is especially true for the consumer electronics used in homes and offices. The main reason is that the air cooling is relatively simple, cost-effective, and safe. Air cooling is also a main cooling method for all current telecommunications equipment. However, due to poor thermal properties of the air, there is a system power limitation for the air cooling.

The thermal technology map for various power systems with different cooling schemes is developed and presented in Figure 10.5 [5]. The horizontal axis shows the estimated heat transfer coefficients. The left vertical axis in the figure gives the system power while the right vertical axis represents the system complexity and cost. Generally, the cooling system becomes more complicated and costly when the system power increases. It should be noted that Figure 10.5 should only be used as a general guideline. It is believed that the upper limit of the forced air cooled rack is about 25,000 W. Any system power greater than the above value,

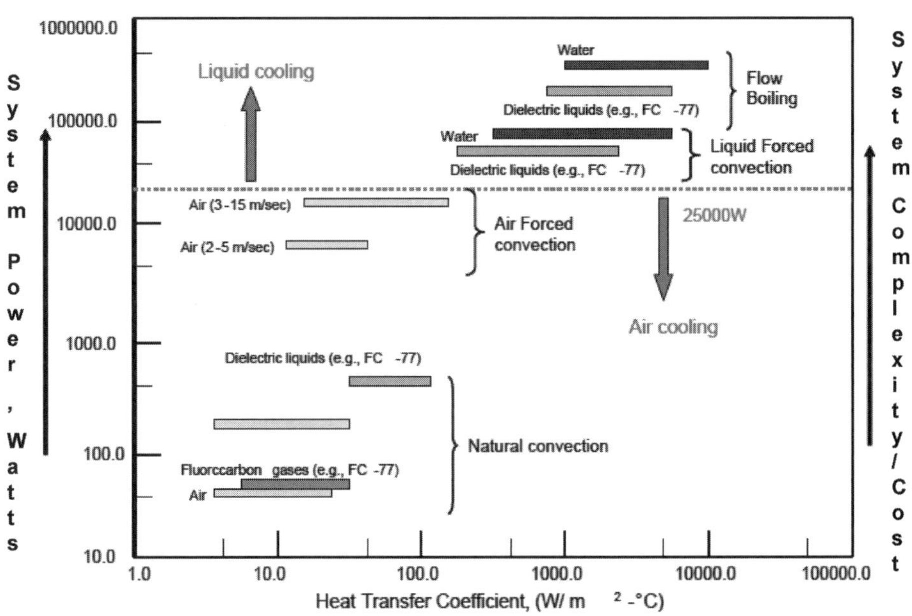

FIGURE 10.5 Thermal technology map.

liquid cooling may be considered. For practical engineering applications, the forced air cooling is not recommended for any system over 20KW even it works thermally because of acoustic noise and the COP issues.

The overall thermal management challenges in the next generation telecommunication equipment are as follows:

1. Heat removal from high density and high heat flux packages within reducing volume with low pressure drop.
2. Hot spot thermal management in packaging level
3. Heat spreading from chip packages (small area) to heat sinks (large area)
4. Thermal integration with EMI for high speed, low voltage packages
5. Integration acoustic noise management into system thermal solutions
6. Precise temperature control for optical/photonic packaging
7. New interface materials with high thermal conductivity and matching CTE
8. Advanced analysis tools for thermal modeling
9. Improvement on PCB thermal performance, i.e, increasing overall thermal conductivities
10. Integration energy efficiency into system thermal design

One must also consider the liquid cooling technologies for the high power telecommunication systems, especially for the system power over 25KW in order to meet thermal and acoustic noise requirements as well as achieving system energy efficiency. Specific thermal management methods in chip and system levels will be discussed later.

10.3.1 Chip Level Thermal Management

The thermal issues in the chip packaging are due to the increase in chip power and the local hot spots. The primary goals of thermal management are to minimize the component junction temperatures and the temperature gradients of the chip packages. The detailed thermal paths within a chip package are given in Figure 10.6, and the explored view of a simplified sketch of the chip packaging and the heat sink are illustrated in Figure 10.7. The thermal resistance, θjc (junction to case) which includes a combination of a series and parallel thermal resistance from the junction (heat source) to the case should be as low as possible. Among all thermal resistance, the TIM1 (thermal interface material) is the most critical factor because it controls the heat transfer from the junctions to the case.

The possible methods of reducing the ΔT across the TIM1 can be described as follows:

a. Thin film thermoelectric device (by Nextreme)
b. Indigo-1 Phase change metallic interface with K = 20–40 W/m °K (by Enerdyne Solution)
c. Reactive bonding of multilayered foils composed of thousands of nanoscale layers (by Reactive Nano Technologies, Inc.)

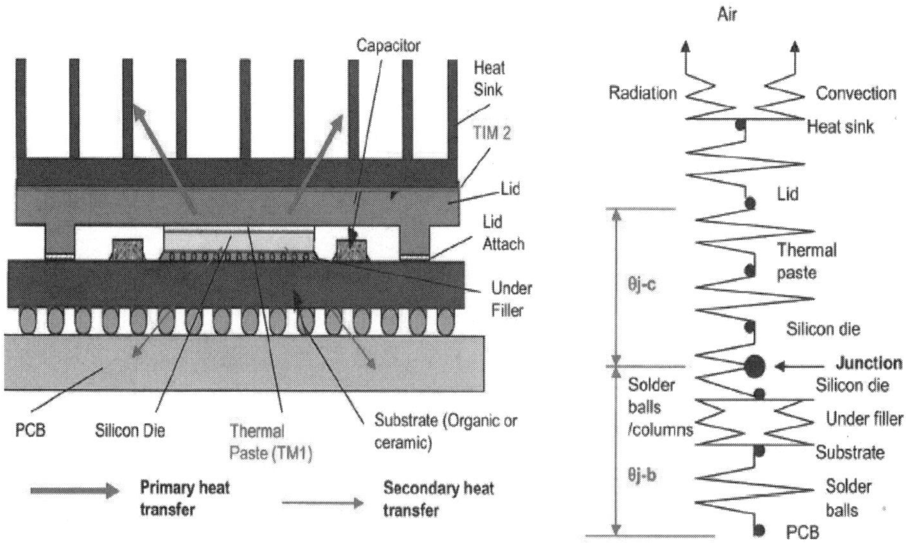

FIGURE 10.6 Thermal paths in chip package with heat sink.

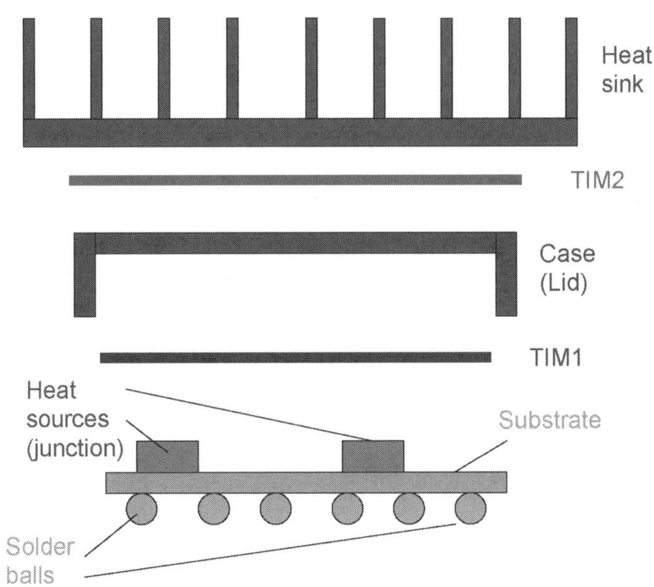

FIGURE 10.7 Sketch of chip packaging and heat sink.

d. Any materials with high thermal conductivity, e.g., Indium with K = 86 W/m °K (by Indium Corporation).

Similarly, the following methods can be adapted to lower the ΔT of the TIM2

a. Indigo-1 phase change metallic interface (by Enerdyne Solution)
b. Thermal grease with high thermal conductivity
c. Thin metallic foils

Generally, heat is generated from metalized areas of the silicon dies inside a package and is transferred to the component case (lid) and also to the printed circuit board. Heat conduction is the primary heat transfer mode inside a component. Cooling by conduction often requires the transfer of heat across various materials and interfaces that may be laminated or bonded together. Therefore, materials used in the packaging should have high thermal conductivity for heat conduction. In addition, the coefficient of thermal expansion (CTE) of materials is as important as the thermal conductivity. This is due to the fact that any mismatching of CTE often results in separation of the joints. Therefore, materials should not only have high thermal conductivity but also have matching CTE.

Because of increasing in component power, several metal-matrix composite materials such as Al/Gr (aluminum/graphite), Cu/Gr (copper/graphite) that have thermal conductivity as high as aluminum or copper, but with the coefficient of thermal expansion close to silicon or gallium arsenide have been developed. The specific value of the CTE can also be obtained by adjusting the percentage of individual elements in the composite. The thermal conductivity and CTE of various materials is presented in Figure 10.8 [6].

Due to the increased power requirement, the vertical integration of the circuit die, which is also referred to as "3D Packaging," has received considerable attention in the recent years. Thermally, the vertically stack chip packages are even more demanding than the traditional 2D chips where heat is generally conducted through the multiple interfaces and layers of the different materials to the case (lid) and to the board in the traditional 2D package. The major challenge in the thermal design of the 3D chip package is how to remove heat from interior layers of the dies. For these high power 3D chips, the traditional air cooling techniques are not possible because of extremely poor thermal properties of the air. It will require unrealistically high airflow rates for effective air cooling. The microchannel water cooling on the integrated circuits was first proposed by Tuckerman and Pease [7, 8]. The backside heat removal using the microchannel cooling technology has been shown effective for the single layer die. The interlayer integrated water cooling proposed by Alfieri et al. [9] as shown in Figure 10.9 is a very promising approach.

One of special type components which are unique to telecommunication systems is the small form-factor pluggable (SFP) transceiver. The SFP as shown in Figure 10.10 is a component used for both telecommunications and data communications applications. Generally, a SFP is inserted into a steel cage with perforation on the top surface as illustrated in Figure 10.10. The cage is soldered to the printed circuit board. The overall dimensions of a SFP are 8.5 mm (H) × 13.4 mm (W) × 56.5 mm

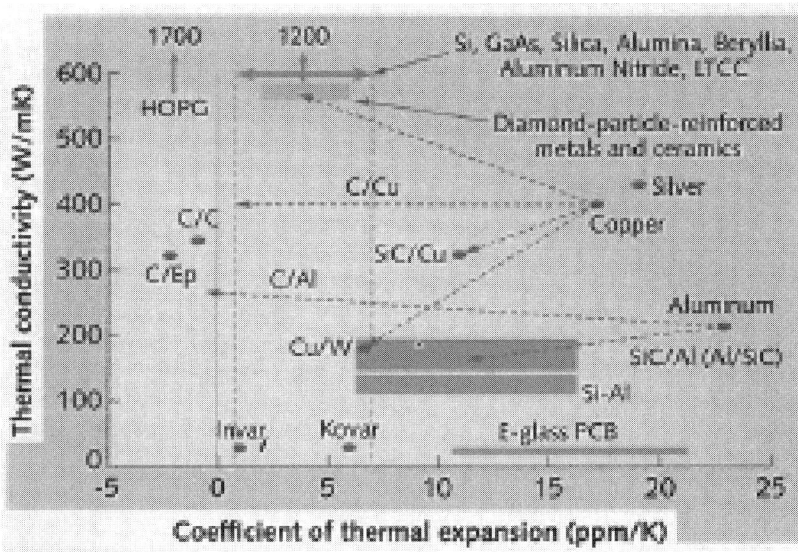

FIGURE 10.8 Thermal conductivity and CTE of various materials.

(L) and its heat dissipation typically ranges from 0.7 to 1.2 watts with the module temperature limit at 70°C. Because the SFP must be freely removed or plugged into its cage, a small air gap (about 0.25 mm) exists on all sides between the SFP module and the cage which results in a very high thermal resistance. With typical inlet or ambient air temperature of 50 to 55°C. It is sometimes difficult to maintain the SFP module case temperature at 70°C, especially under natural convection environment. Therefore, a heat sink is often needed (Figure 10.11).

Another product from SFP family is XFP which is larger than SFP in size. The overall dimensions of a typical XFP are 8.5 mm (H) × 18.35 mm (width) × 78.0

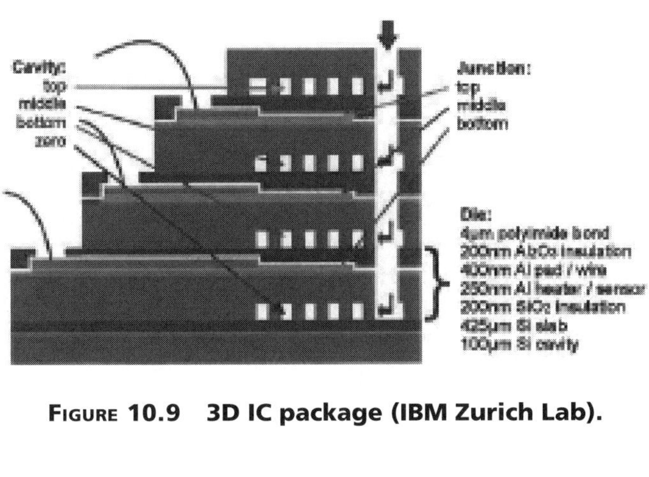

FIGURE 10.9 3D IC package (IBM Zurich Lab).

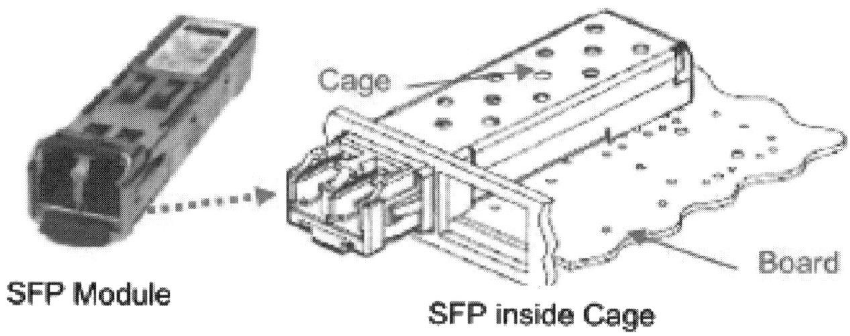

FIGURE 10.10 SFP module/cage on board.

mm (L). The XFP power can be up to 3.5 W with the temperature limit of 70°C. Actually, the power of some of the newly developed SFP can reach up to 3.5 W too and makes thermal design ever more challenge and difficulty. Because of the air gap (about 0.25 mm) between the module and cage along with the low allowable temperature limit (70°C), SFP or XFP becomes one of most critical components, especially for 3.5 W components on the board thermally. Lee and Lodhia [10] performed thermal analysis and testing on 1.5 W XFP at air various velocities ranging from 1.016, 1.778, and 2.54 m/sec. However, their work was mainly aiming at the development of thermal models.

To overcome the high thermal resistance of the air gap, a large rectangular hole is cut on the top surface of the cage so that the heat sink base can be in contact with the SFP module as shown in Figure 10.12. The bottom piece with trapezoid shape which is an integrated part of the heat sink base will make a direct contact with the SFP module. Therefore, the thermal resistance between the heat sink

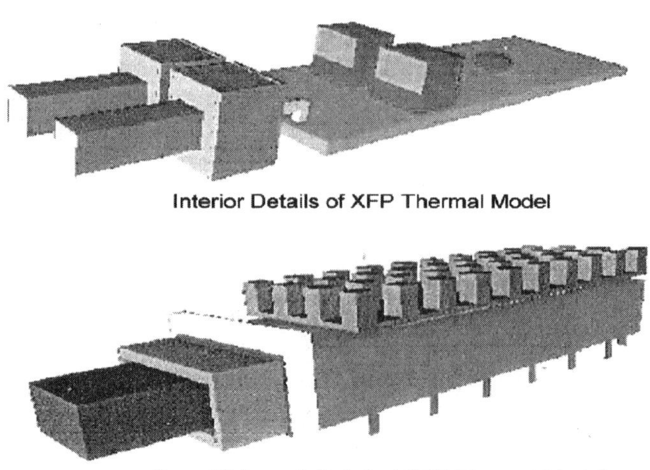

FIGURE 10.11 Thermal models for XFP.

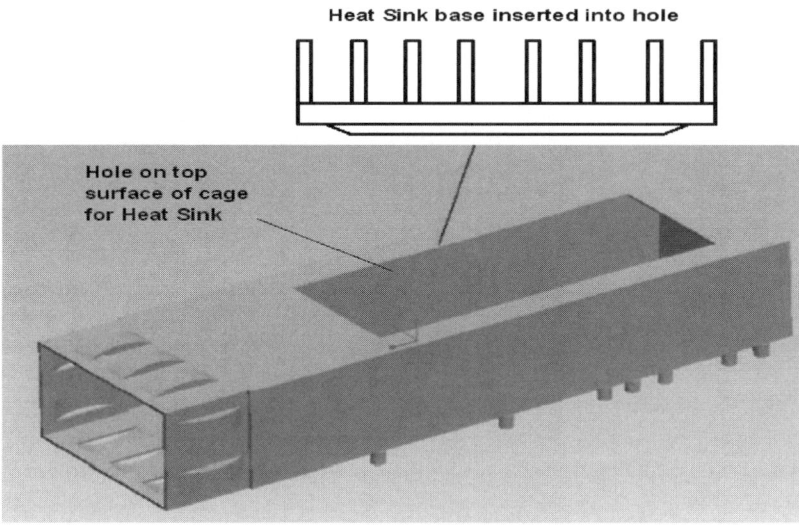

Figure 10.12 Heat sink and SFP/XFP cage.

and the module can be reduced. The compounded thermal problems are due to a series of SFP/XFP in a row on a given board as shown in Figure 10.13. The number of modules in a row may be more than 20. Because of preheating of air stream, the modules at far downstream are very difficult to meet the temperature limit of 70°C. Large heat sink which serves multiple components by a single heat sink as shown in Figure 10.12 is recommended in order to reduce the maximum temperature of the modules at the downstream of the airflow path. For module power greater than 2.5 W, especially at 3.5 W, it is sometimes impossible for these modules at the downstream of the airflow path to meet the temperature limit of 70°C with the inlet air or ambient air over 50°C. Therefore, an industry grade SFP/XFP whose temperature limit is at 85°C is required. In addition, thermoelectric coolers can also be employed to resolve the thermal issues of these components. However, thermoelectric cooler would require additional electric power.

10.3.2 System Level Thermal Management

Among the many system level challenges faced in telecommunication equipment manufacturers, the power density is the most critical one thermally. Extrapolating from current systems, future nodes would dissipate hundreds of kilowatts for power and the power trends in the projected growth are expected to be continuously increased as previously presented in Figure 10.4. The power reduction strategies and high-density thermal management become two of the critical elements for the future telecommunication equipment. The thermal technology map, as shown in Figure 10.5, can be used to select the proper cooling schemes.

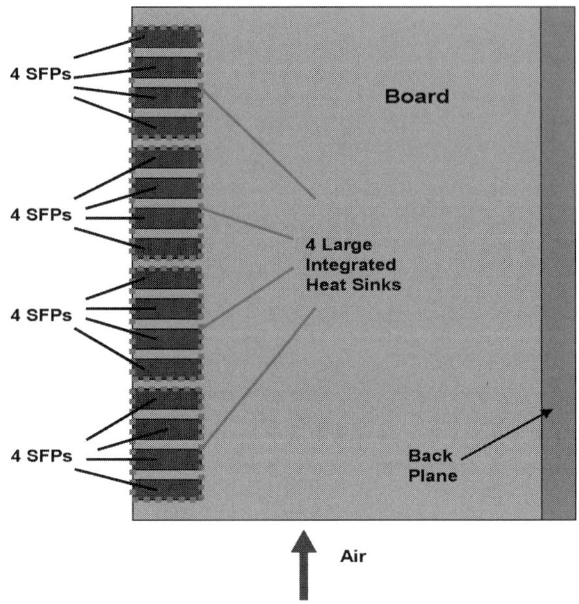

FIGURE 10.13 Example of series of SFP modules on board.

Air cooling has been and is still the key element for the thermal management of the telecommunication equipment. To achieve the efficient thermal management schemes in air cooling, the following factors should be considered

1. Optimization of heat sinks on components
2. Thermal interface materials (TIMs) with high thermal conductivity
3. Good airflow management with lower system pressure drop and better flow distribution over the boards

The purpose of using a heat sink is to enhance the heat transfer from the component to the air stream. The typical extrusion heat sinks may not be suitable for the component power greater than 35 W. The main reason is the limited available heat transfer surface. This is due to the fact that the extrusion fin heat sinks are bulky and heavy with thick fins. The other types of fin structures such as the plain fins may be considered. Another issue to deal with is the base spreading resistance of the oversize heat sinks. To reduce the thermal spreading resistance, the common methods are to increase the lateral heat conduction at the heat sink base. The approaches include, but are not limited to, the embedded graphite spreader (or any high thermal conductivity materials), embedded heat pipes, vapor chamber, or liquid chamber base.

The flow management is another important factor for achieving the effective thermal management scheme. The objective of the flow management is to obtain proper flow distribution over the boards and also to minimize the system flow resistance or pressure drop in order to increase the system flow. The baffles are also often employed to improve the flow distribution over the board or to guide flow to where it is needed. In addition, system design must have a good balance between the desired system thermal performance and the pressure drop. In order to reduce the acoustic noise and also to achieve system energy efficiency, i.e., high COP, the following guideline for the system pressure drop versus flow rates as presented in Figure 10.14 is recommended for air cooled systems.

Liquid has often been used for cooling of the high power systems. The greatest advantage of liquid cooling over the air cooling is its cooling capability. Though the liquid cooling has long been employed in the high-performance computers, the current practice of using liquid (i.e., water) at the equipment in the data centers is aiming at the reduction of the air temperature to the data centers. IBM System 360 Model 91 in 1967 was the first IBM computer to use water-cooling in any form. The water-to-air heat exchangers as shown in Figure 10.15 [11] are installed at the door and are inserted between the card cages when the door is closed. The purpose of the water-to-air heat exchangers is to reduce the inlet air temperature to the upper card cages (boards) in downstream of the airflow. As can be seen from the figure, the ΔT air is reduced by more than 20°C with the heat exchangers in the rack.

On the other hand, the IBM RDHX (rear door heat exchanger) [11] which is a water-to-air heat exchanger is attached to the exits of the server rack as illustrated in Figure 10.16. The overall dimensions are 25 in. (W) × 5.6 in. (D) × 76.4 in. (H). The main purpose of the RDHX is to lower the temperature of the exit air from the servers to the date center so that the energy consumption of the air-conditioning

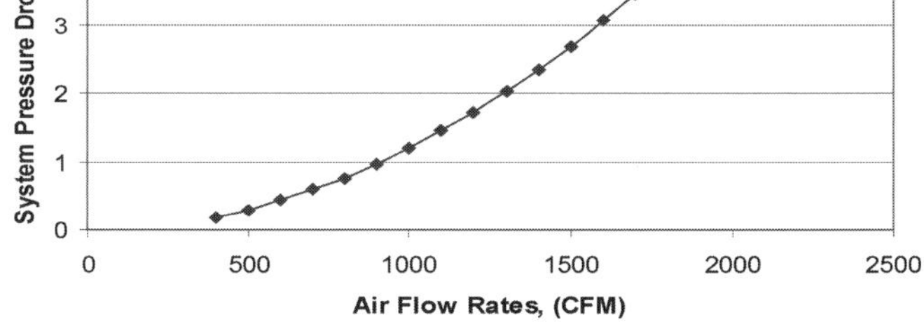

FIGURE 10.14 Recommended system maximum pressure drop versus flow rates.

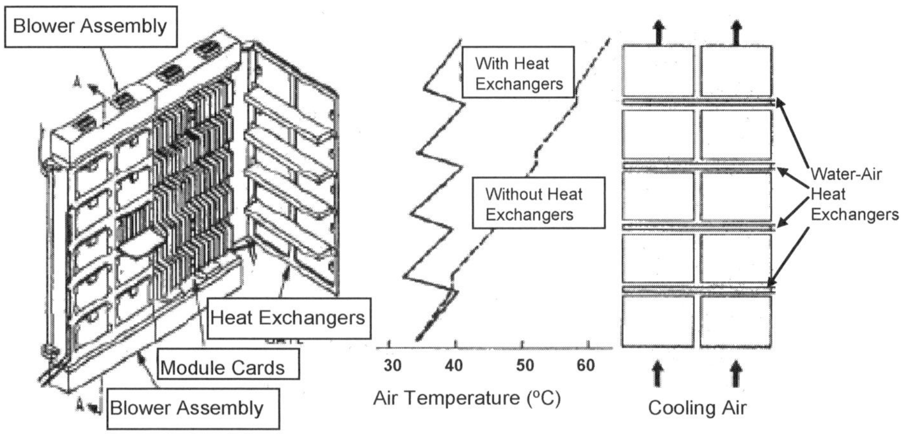

FIGURE 10.15 IBM water–air heat exchangers in cabinets.

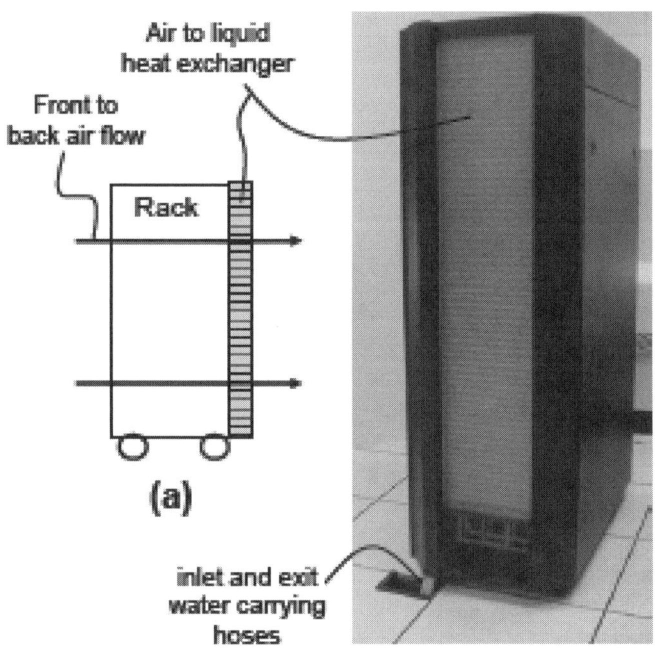

FIGURE 10.16 IBM rear door heat exchanger.

units in the data center can be reduced. It should be noted that the RDHX does not affect the thermal performance of the equipment itself.

The hybrid cooling of the supercomputers [11, 12] has recently received considerable attention. In other words, the high-power components on the board are cooled by the liquid, while the rest components are still cooled by the air. The issue of possible condensation inside the rack must carefully be considered. The possible fan failure and also the acoustic noise issue may still exist in the hybrid cooling scheme. In addition, the liquid loop (tubes) may become an obstacle of the certain airflow to the components. Recently, IBM Zurich Research Lab in conjunction with ETH developed a warm water (60°C) cooled supercomputer (Aquasar) [13, 14] using microchannel cold plate for cooling the components. The active liquid cooling (total liquid cooling) is the best approach to achieve the energy efficiency both at equipment level and data center and also to support the much higher system power while eliminating the acoustic noise from air cooling. The actively liquid cooled rack has following advantages over the current air cooled rack.

- Liquid cooling can support much higher system power
- Liquid cooled rack makes equipment much more reliable
  - By eliminating fans, equipment reliability is significantly increased
  - Effective liquid cooling lowers component temperatures
- Liquid cooled rack is much compact and small
  - By eliminating fan tray, air filter, air inlet and exit spaces
  - Slot size (spacing between boards) is small
- Equipment is almost maintenance free (no need to replace air filter and fans)
  - Operation cost is also reduced by eliminating fans and air filters
- Liquid cooled rack can totally eliminate acoustic noise from fans and high speed airflow
- Liquid cooled rack is independent of ambient conditions (e.g., room air temperature) as long as chilled water is available
- Eliminating hot air from equipment will reduce data center temperature and decrease energy consumption in data centers

10.4 THERMAL DESIGN OF LIQUID COOLED TELECOM EQUIPMENT

The liquid cooling has not been adapted commercially in any telecommunication equipment. However, due to the increased system power and also the concern about the energy efficiency, the liquid cooling has received considerable attention in telecommunications industry recently. Alcatel-Lucent [15] has developed a sealed cabinet using dielectric mist for cooling as illustrated in Figure 10.17.

The large droplets of dielectric mist from the atomizer are injected into the upstream of airflow entering the printed circuit boards. The liquid mist is heated by the electronics on the board and is evaporated along the air stream. In other

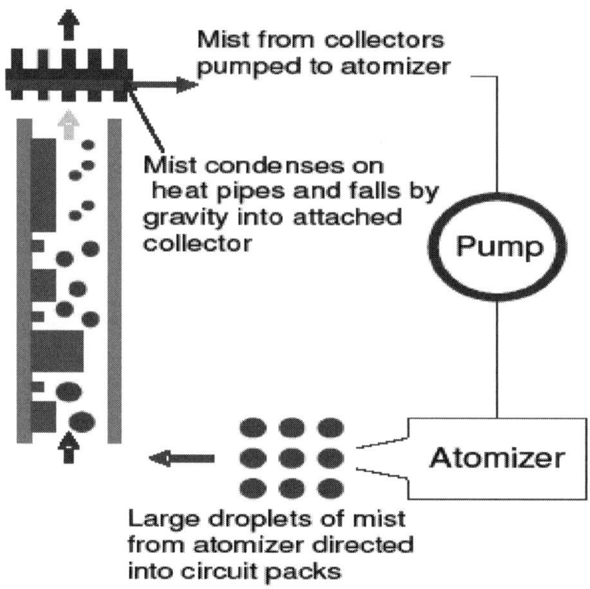

FIGURE 10.17 Alcatel-lucent enhanced cooling in a sealed cabinet.

words, heat generated from electronic components is to vaporize the dielectric mist instead of heating up the air stream. The mixture of air and the dielectric vapor flows over an array of heat pipes and the dielectric vapor is then condensed and the liquid is fallen into the liquid collector and finally is pumped into the atomizer for reuse.

Cisco also employed the hybrid cooling for the high power systems which are the modified 7609 Series [16]. The back side of chassis, including the heat exchangers, tubing and the pumps is shown Figure 10.18. The system under consideration is a standard 19 in. rack which consists of 9 boards with the board pitch of 1.8 in. The maximum system power per shelf is about 10 KW with the maximum nominal power per board of 1000 W.

Both systems from Cisco and Alcatel-Lucent are still in the very preliminary phase in the laboratory. Some of other telecommunication equipment manufacturers have also started working on the liquid cooled systems.

The chilled water is readily available in most of the data centers. The sketch of the active liquid cooling can be depicted as illustrated in Figures 10.19 and 10.20 [5]. Figure 10.19 shows a general liquid cooling loop in the equipment. The heat exchanger as shown in the figure can be the cooling towers or condensers of the data center. The cooling concept of the active liquid cooled rack is given in Figure 10.20. The liquid enters the inlet manifold which provides proper flow rate to the individual boards and leaves the equipment through exit manifold. The chilled water in Figure 10.20 is directly taken from the data center. It should be noted that the pump may not be needed in this case if water from the date center has

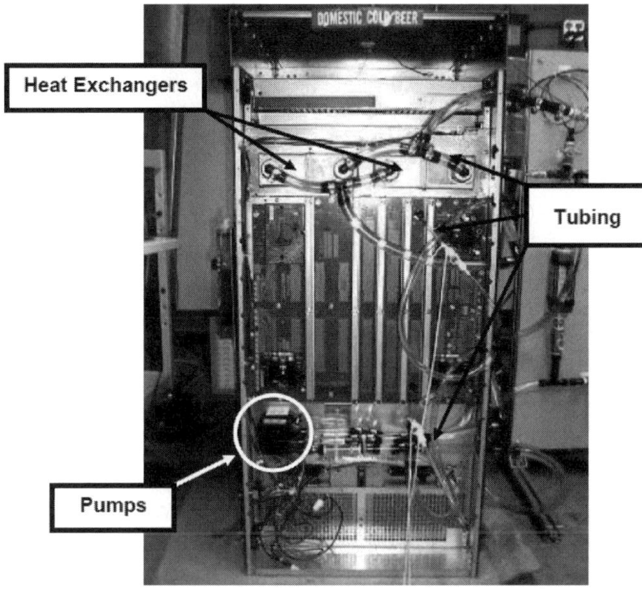

Figure 10.18 Cisco liquid cooled rack : back of chassis exposed.

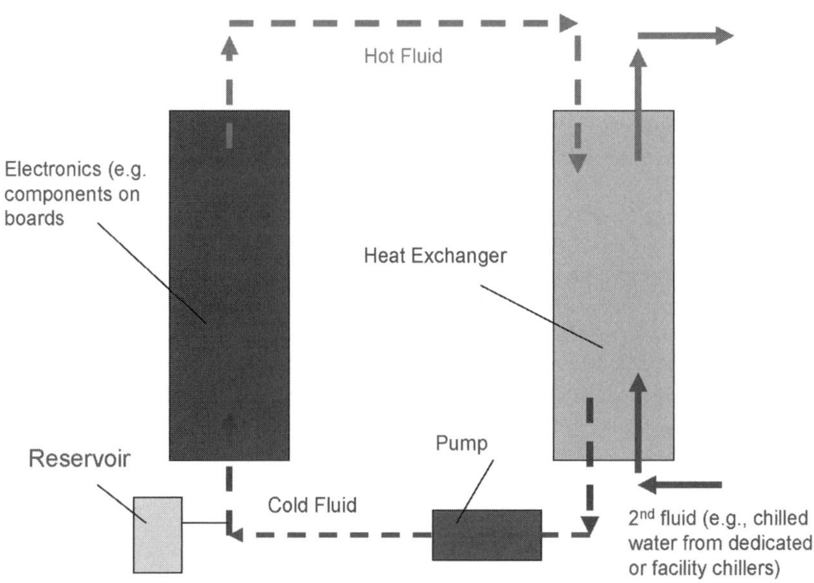

Figure 10.19 General liquid cooling loop in equipment.

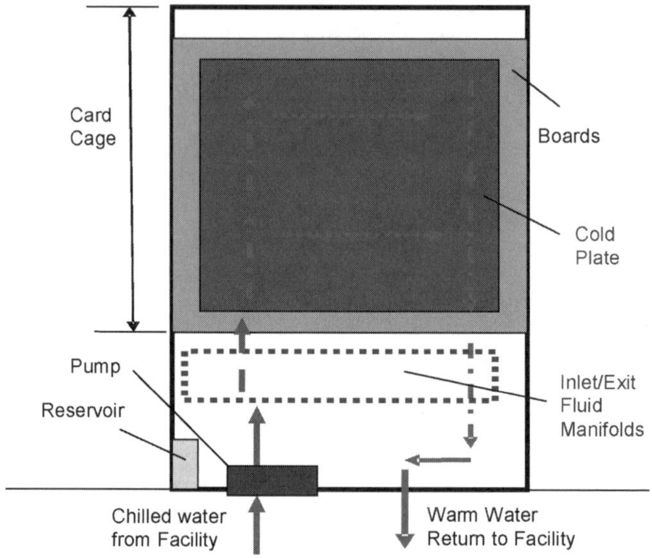

FIGURE 10.20 Active liquid cooled rack with facility water.

sufficiently high pressure. The flow rate to individual boards must be regulated in proportion to the power of the boards.

Figure 10.20 shows the liquid cooled equipment with the facility chilled water. On the other hand, Figure 10.21 [5] illustrates the liquid cooled equipment with the self supplied liquids. In this case, an additional heat exchanger is required in order to cool the hot liquid from the equipment to be reused again. This heat exchanger can be a liquid-to-liquid heat exchanger or an air cooled condenser.

Figure 10.21 can also be applied to the outdoor systems wherein dielectric liquids must be used in the outdoor applications. This is due to the fact that the outdoor equipment is typically required to operate at the temperature below –40°C or even –50°C. In general, a dielectric liquid which has a high dielectric strength to prevent electrical breakdown is a preferred choice in cooling of electronics. Unfortunately, the heat transfer capacity of the dielectric fluids is much lower than that of the water. The dielectric liquid is not normally required for the cases with indirect liquid cooling. However, it is a must for the immersion cooling where liquid is in direct contact with electronics.

The board assembly, including board and liquid network must be able freely to be removed or plugged into the equipment. To do so, each board requires at least two quick disconnects for the liquid loop. A good quick disconnect which is extremely expansive is needed to avoid any liquid leakage after the board assembly is unplugged. Figure 10.22 gives a sketch to illustrate the connection of the board assembly to the system. It should be noted while the board is plugged into the back plane for the electrical connection and, at the same time, the liquid loop must be

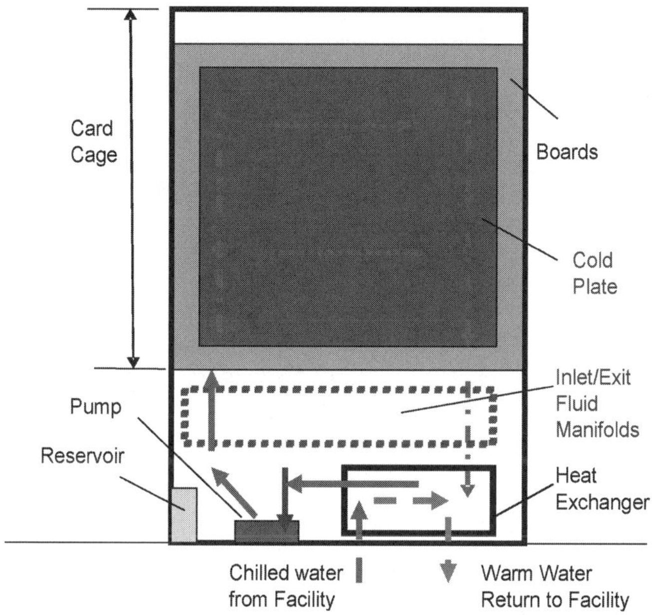

FIGURE 10.21 Active liquid cooled rack with self supplied coolant.

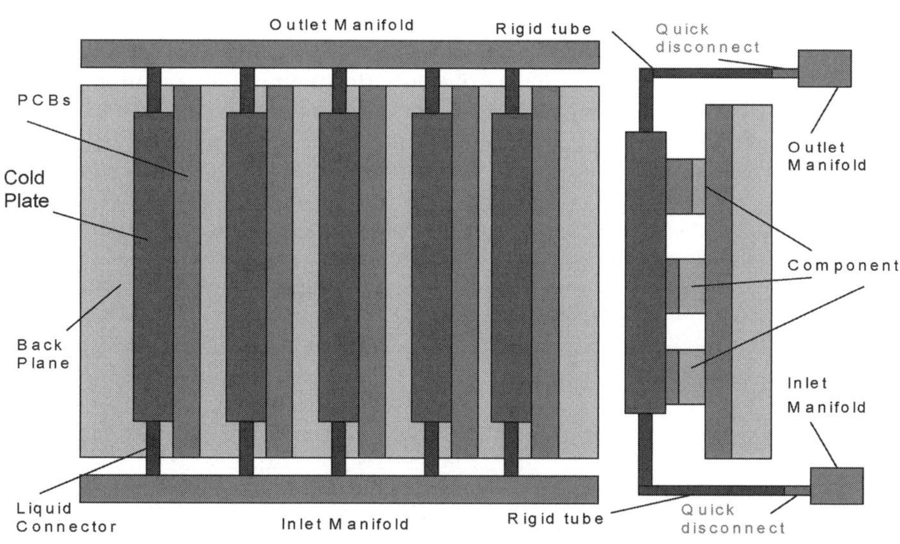

FIGURE 10.22 Sketch of board assembly connection to system.

connected to the system too. The fluid connections between liquid network and the liquid inlet and exit manifolds in the system are made through the quick disconnect fittings. The figure, which shows one cold plate for all components on the board, is just for illustration only. Due to various heights of the components on the board, multiple cold plates may be required.

Water is the most common fluid used in the high power computers because of its high heat transfer capability. However, water cannot be used in a direct immersion cooling nor for outdoor applications. The former is due to its poor dielectric properties and the later is because of its high freezing temperature. When water is used, it also must be properly treated for anti-corrosive and anti-fungal. For material compatibility, copper or stainless steel must be used for the entire liquid loop if water is adopted as the coolant. On the other hand, aluminum can be employed for the dielectric fluids. However, stainless steel is recommended if de-ionized water is the coolant. All above three types of materials (copper, aluminum, and stainless steel) are compatible with Glycois.

10.5 ENERGY EFFICIENCY

As stated previously, the equipment power consumption is continuously increased at a fast pace. In addition, the cost of powering and cooling in the equipment at the data centers has also increased significantly in the last few years. Therefore, the energy efficiency at the equipment level and the data center become a top priority in the industry.

Figure 10.23 [17] illustrates the power consumption in a typical data center. Among them, the power consumption for the IT and cooling equipment are 45.4% and 30%, respectively. Another example for the 1 U servers, the ratio of the 3-year site cost to the 1 U server cost is given in Figure 10.24 [18]. As can be seen from the figure, the 3-year cost of powering and cooling servers is about 1.5 times of the cost of the servers in 2006, and this factor is projected to be over 22 times at the worst case by 2012.

In addition, the power consumption of equipment has a great impact on the power required in a data center which is directly related to the operation cost. Based on the estimate by Verizon [19], 1 W saved on the equipment level will save a total of 2.41 W in their respective the data center as shown in Figure 10.25. Therefore, it is extremely important for the equipment makers to do a better job on energy efficiency of their products. It should be noted that this cascade factor varies slightly with the individual data centers.

The following steps [5] should be considered to achieve the energy efficiency at the equipment level:

1. Using low power consumption components, especially for power supply.
2. Utilizing industry grade components whenever possible. The industry grade parts are more expensive but are more reliable. In addition, the industry grade component can withstand higher temperature that in turn requires less airflow.

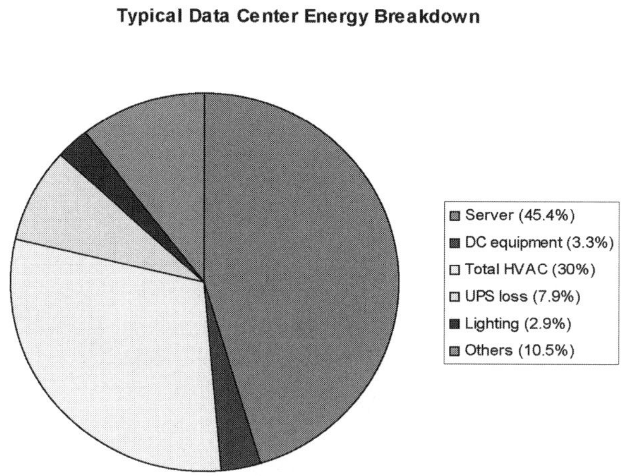

FIGURE 10.23 Typical data center energy breakdown.

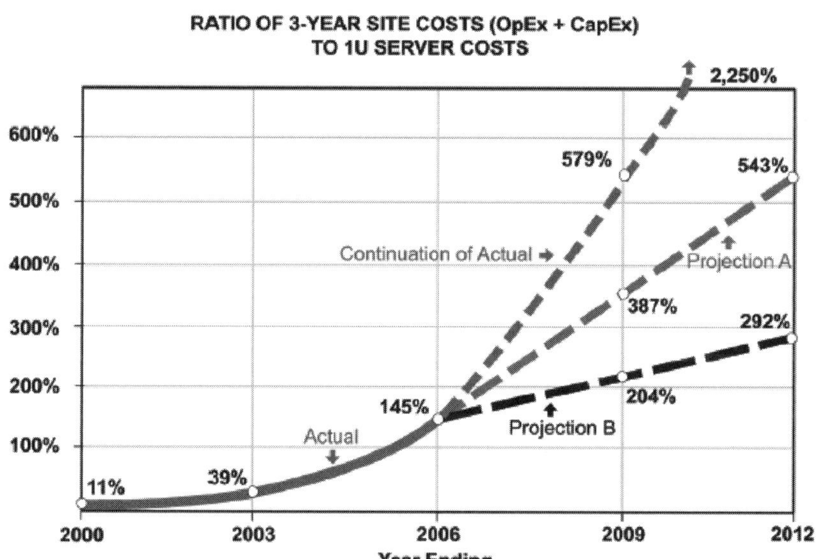

FIGURE 10.24 Ratio of 3-year site cost to 1 U server costs.

3. Employing efficient thermal control schemes. The fans can operate at low speeds because the less flow is required with effective heat transfer methods.
4. Designing equipment with power management features. With this feature, some of boards can shift to standby mode in the conditions of low traffics or demands.

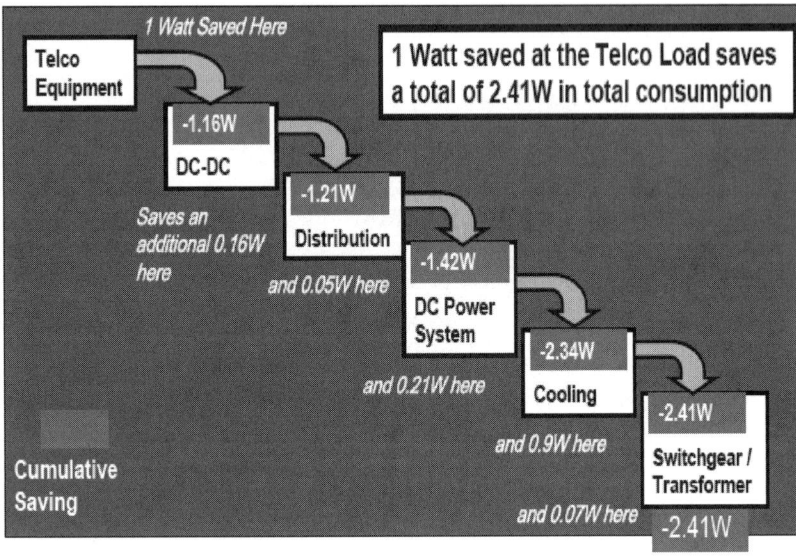

FIGURE 10.25 Cascade effect of energy savings from equipment to data center.

5. Adapting passive cooling methods whenever possible. The passive cooling is simple, reliable, cost effective and energy efficient but is limited to the low power systems due to poor thermal properties of air.
6. Adapting liquid cooling, especially the actively liquid cooled equipment.

In addition, one should minimize the total power per network throughput by transferring bits in the optical domain from the network architectural point of view. This so-called all-optical network is a less power hungry network.

10.6 ALTERNATIVE ENERGY

With significant increase in the energy consumption and cost in the past few years, it becomes necessary to explore various types of alternative energies to replace or to supplement the energy consumptions in the equipment as well as in the data centers. Among them, the thermoelectric power generation and the fuel cell technology power generation have been received considerable attention recently.

10.6.1 Thermoelectric Power Generations

Two major classes of application of thermoelectric devices are heat pumps and power generators. In the former, electrical energy is supplied to transport thermal energy from one location to other locations, while in the latter, thermal energy is converted into electrical energy. The power generation from the waste heat of the electronic systems is the subject area to be discussed here.

The basic theory of the thermoelectric device is due to the Seebeck effect which states an electromotive force (emf) or voltage will be generated if two junctions of two different materials are held at different temperatures. This emf is proportional to the temperature difference between two junctions as follows

$$dV = \alpha_{12} dT \quad (10.1)$$

and

$$\alpha_{12} = \alpha_1 - \alpha_2 \quad (10.2)$$

where αi is Seebeck coefficient for material i with unit of V/°C.

For convenience, the figure of merit, Z which is a measurement of the usefulness of a material in a thermoelectric device is defined as follows

$$Z = \alpha^2/(\sigma K) \quad (10.3)$$

where σ is the electrical resistivity (ohm-cm), and K is the thermal conductivity (W/m °C).

The sketch of a thermoelectric power generation is given in Figure 10.26 [20]. Generally, the higher temperature difference between the hot and cold sides, the higher thermal efficiency will be. The thermal efficiency is limited by the material development. Two critical factors that dictate power output in a thermoelectric power generator are as follows:

1. The amount of heat flux that can be successfully transferred through the module, and
2. the temperature difference between the hot and cold side.

Due to relatively small air temperature difference, typically less than 20°C, between the inlet and exit in telecommunication equipment, the thermal efficiency

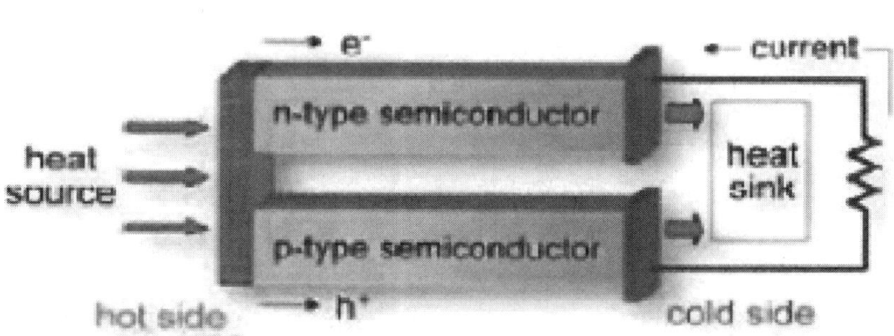

FIGURE 10.26 Basic sketch of thermoelectric power generator.

of the thermoelectric power generator is extremely small. The thermal efficiency with ΔT of 20°C is estimated to be about 2% in 2010 and approaching 5% by 2013.

Figure 10.27 [20] shows one of possible schemes that the waste heat from equipment at data center can be converted to the electricity by the thermoelectric power generators. Even with very small thermal efficiency, it is still worthy to try to utilize the waste heat from telecom or computing systems for the energy generation in the data center because an extremely high amount of heat is generated. The following example is to demonstrate the possibility of using waste heat recovery to generate electric power in order to supplement the power consumption in the data center.

Example 10.1

Assume 5% thermal efficiency with ΔT of 20°C for a 20 KW telecommunication equipment. As stated previously in Figure 10.25, 1 W saved in telecom equipment will save 2.41 W in the data center.

Solution

Savings on equipment level for 20 KW system:

$$\text{Annual energy saving} = 20\ \text{KW} \times 0.05 \times (24 \times 365) = 8760\ \text{KW-HR}$$

$$\text{Annual cost saving} = 8760\ \text{KW-HR} \times \$0.11 = \$963.6$$

Savings on data center for 20 KW system:

$$\text{Annual energy saving} = 2.41 \times 8760\ \text{KW-HR} = 21111.6\ \text{KW-HR}$$

$$\text{Annual cost saving} = 21111.6\ \text{KW-HR} \times \$0.11 = \$2322.3$$

The above calculations are based on the average residential rate of \$0.11/KW-HR.

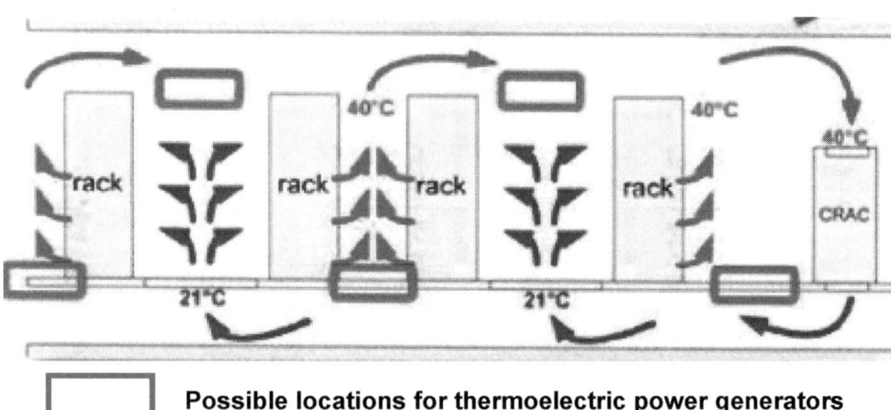

FIGURE 10.27 Possible locations of thermoelectric power generators in data center.

10.6.2 Fuel Cell Technology for Power Generation

The fuel cell is an electrochemical conversion device which was first discovered by Christian Friedrich Sconbein in 1838. It produces electricity from fuel (on the anode side) and an oxidant (on the cathode side) which react in the presence of an electrolyte. The reactants flow into the cell and the reaction products flow out of it.

The basic principal of the fuel cell technology is as follows:

$$H_2 + 0.5\ O_2 \rightarrow \text{Electricity} + \text{Heat} + \text{Water}$$

Due to electrochemical actions, the hydrogen fuel and oxygen produce electricity and heat with the byproduct of water. The hydrogen is the primary fuel source for the fuel cell. The process of fuel reforming allows for the extraction of hydrogen from many sources such as natural gas and propane or any other hydrogen-containing fuels. A typical fuel cell can produce a voltage up to 1V at the full rated load. Therefore, in order to deliver the desired amount of energy, the fuel cells can be combined in series or in parallel circuits, where series yield higher voltage and parallel allows a stronger current to be drawn. Such a design by combining individual fuel cells is called a fuel cell stack. To produce electrical power for a working application requires more than just one fuel cell stack. A fuel cell system may include fuel cell processing, thermal management, water management, power conditioning, electric grid connection, and energy storage modules.

FIGURE 10.28 Direct application of fuel cell energy to OSP operation.

Fuel cell systems offer the potential for clean, reliable energy generation and have been employed to supplement the energy consumption in the data centers. Fuel cells have recently been adapted as the back energy to or even to replace the batteries in the outside plants (OSPs). Plug Power's [21] developed GeneCore 5 KW products to be used as an emergency backup power over 72 hours or more for the cell tower customers. Furthermore, Clear Edge has developed several fuel cell systems to power home and/or business buildings by utilized available natural gas. With OSP moving closer to the residential area where natural gas is readily available, direct application of fuel cell systems to power OSP as shown in Figure 10.28 becomes feasible.

10.7 FUTURE RESEARCH NEEDS

The continued development of new and improved thermal management technologies will require the combined efforts of industry based development and university based research with a focus at practical applications. Extensive heat transfer, thermo-fluid, and thermo-mechanical research are needed to define new opportunities and to improve predictability and reliability.

The specific research needs are recommended as follows:

1. Thermal spreaders
   - Inexpensive, light weight, high thermal conductivity, anisotropic materials (possibly composites) offering a closer thermal expansion match to the CTE of Silicon (Si) or Gallium Arsenide (GaAs) or other die materials.
   - Techniques for achieving improved thermal spreading within a chip to alleviate hot spots due to localized high heat fluxes.
   - High capacity small heat pipes, including micro heat pipes for embedding to heat sink base or high performance vapor chambers in order to reducing base thermal spreading resistance.
2. Thermal Interface Materials (TIMs)
   - Thermal pads, greases, pastes, epoxies, and elastomers load with high thermal conductivity nanoparticles.
     a. Overall thermal conductivity of TIMs should be greater than 15 W/m °K.
     b. Compressibility of thermal pad must be greater than 50%.
   - Innovative new interface materials, including natural graphite base materials.
   - Carbon nanotube (CNT) base technology for TIM applications and bumps to substrates as shown in Figure 10.29 for Fujitsu 100 W class amplifiers in the mobile base stations.
   - Novel techniques/materials to minimize interfacial stresses and thermal resistance.
   - Correlations and analytical relations to predict fatigue life of bonded interfaces.
   - Void free processes for thermal pastes and epoxies bonding.

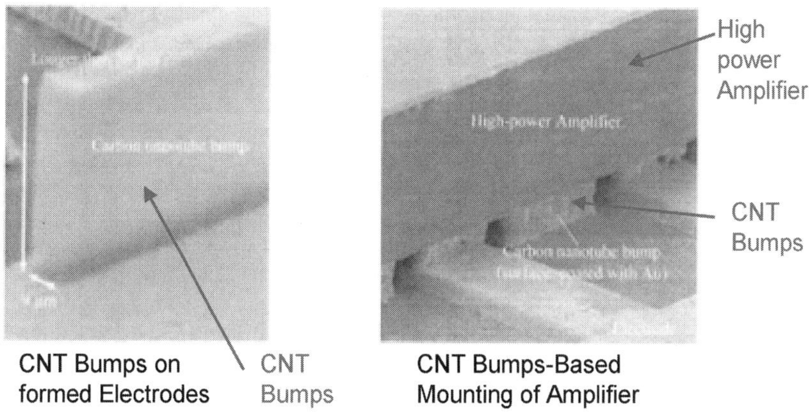

FIGURE 10.29 Carbon nanotubes bump on electrodes and substrate.

3. Air cooling
   - High performance, low power consumption, and low acoustic noise fans or blowers.
   - High efficiency analysis tools
     a. Models and correlations to accurately predict heat transfer in transition and perturbed low Reynolds number flow over packages and in heat sink passages.
   - Heat sink design and optimization for minimizing heat sink thermal resistance subject to mass and volume constraints.
   - Advanced manufacturing techniques for light weight metal or composite material heat sinks.
   - Novel miniature synthetic jet fans for enhancing hot spot cooling.
   - Novel 3D microstructure air filters to enhance product life.
     a. Current air filters collecting dusts and dirt on the front surface which leads to much shorter life.
   - Special subjects
     a. Large heat sink for group of components

     As illustrated in Figure 10.13, a large heat sink is utilized to serve a group of components. This design concept has recently received considerable interests, especially on the high power boards. An oversize heat sink which is generally required for any high power component requires 4 mounting screws attaching to the board. The heat sink mounting is becoming a major issue due to the following reasons:

     i. Due to increased component power density, the number of components required heat sink is significantly increased. On the other hand, the available space for heat sink mounting is limited because of a large number of components on board.
     ii. Holes on board required for heat sink mounting will make board wire routing extremely complicated and difficult.

Therefore, the concept of using a large single heat sink to serve a group of components becomes necessary for high power boards in order to reduce the number of screws needed for mounting heat sinks.

b. Synthetic jets
Using synthetic jets for cooling of electronics has received a great deal of interest in recent years. A synthetic jet flow is a type of flow which is made up of the surrounding fluid. Synthetic jet flow can be developed in a number of ways such as an electromagnetic driver, a piezoelectric driver, or even a mechanical driver, e.g., piston. Impinging jets are a well established technique for achieving high local convective heat transfer rates. Therefore, synthetic jets are employed to enhance local heat transfer coefficient. It should be noted that the flow rate of a synthetic jet which is generated from the surrounding fluid is relatively small. The heat transfer enhancement is due to high velocity jet action.

The discussion here will be focused at the synthetic jets generated by an actuator and a piezoelectric fan.

i. Piezoelectric fan synthetic jets
Piezoelectric fans are vibrating beams, disks or plates whose vibratory motion is actuated using a piezoelectric material. The basic theory is that an ultra low power alternating currents input to the piezoelectric element forces oscillations throughout the beam which can be used to induce flow in surrounding fluid and provide heat transfer enhancement. As these devices (piezoelectric fans) typically run at resonance, significant vibration amplitudes are achieved with small amount of input power. In addition, these devices produce little or no audible noise. The pictures of a typical piezoelectric fan and its application to a heat sink are shown in Figure 10.30.

As described by Kimber and Garimella [22, 23], thermal performance of a piezoelectric fan may be affected by a total of five variable parameters such as frequency, amplitude, width, length and gap of piezoelectric fan. Key physical insights and results from their work include the following:

- The maximum performance for a given fan can be described in terms of the vibration frequency and oscillation amplitude only.
- The frequency is found to be somewhat more influential than the amplitude in determining the largest attainable heat transfer rates.
- The width, length, and frequency, each exhibit the same influence on the value of optimum gap and the exponential decay rate from the peak performance.

The comparison of the heat transfer coefficients of piezoelectric fan and those other types of heat transfer is presented in Figure 10.31 [24].

Experimental investigation of the thermal performance of piezoelectric fans is presented by Acikalin et al. [25]. Prototypes of the fans were built and tested to assess their feasibility and cooling performance and determine optimal locations for the fans. An enclosure the size of a cellular phone and commercially available laptop computer were used to demonstrate the cooling feasibility of the fans. Piezoelectric fans were found to offer significantly localized cooling, exceeding

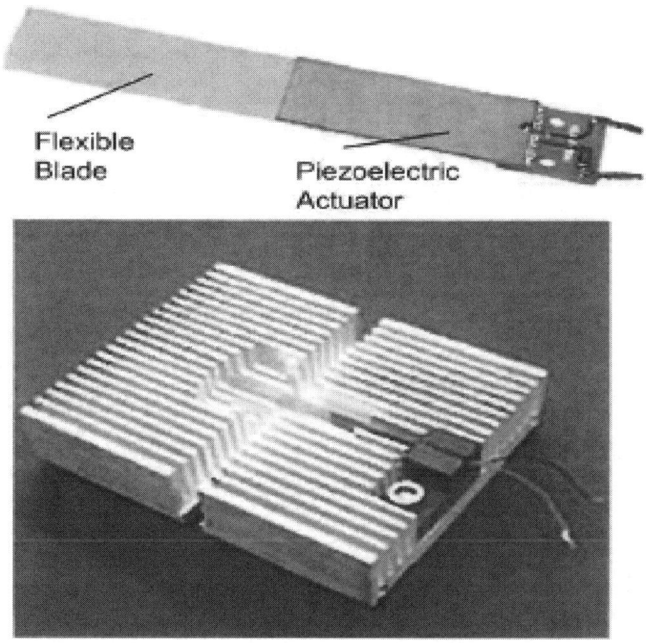

FIGURE 10.30 Piezoelectric fan and its application to heat sink.

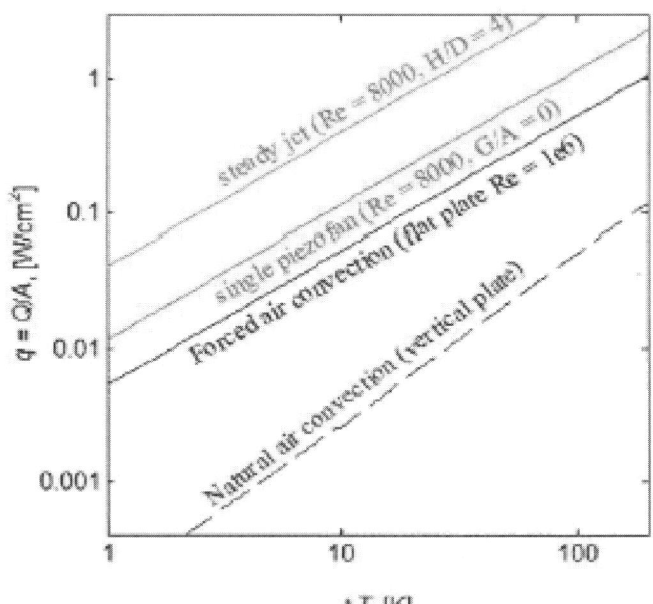

FIGURE 10.31 Thermal performance comparison with piezoelectric fan.

enhancements in convective heat transfer coefficient of 100% while exhibiting low power consumption, minimal noise and small dimensions. Additional work by the authors is given by Acikalin and Garimella [26]. The heat transfer enhancement by piezoelectric fans is limited to the localized cooling. Therefore, the piezoelectric fans have never intended to replace the conventional fans. The best applications of such fans are in the low power systems, especially in the low profile boxes under the natural convection environments. Acikahn, Sauciuc and Garimella [27] studied the piezoelectric fan for low-form-factor electronics cooling. The schematic of two experimental orientations is shown in Figure 10.32. The cases with the best thermal performance in each orientation are given in Figure 10.33. As can be seen from the figure, the thermal performance of the horizontal

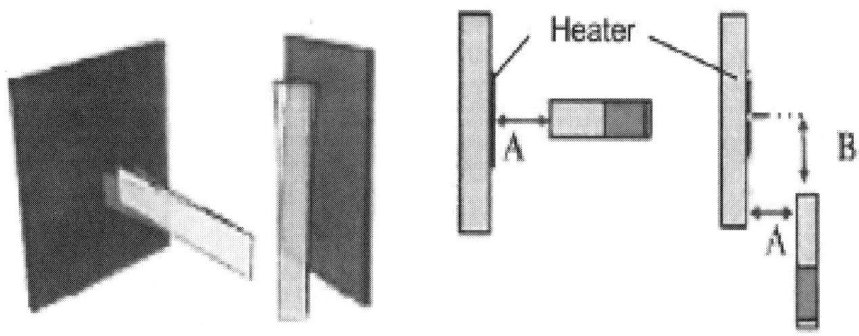

FIGURE 10.32 Schematic of two experimental orientations.

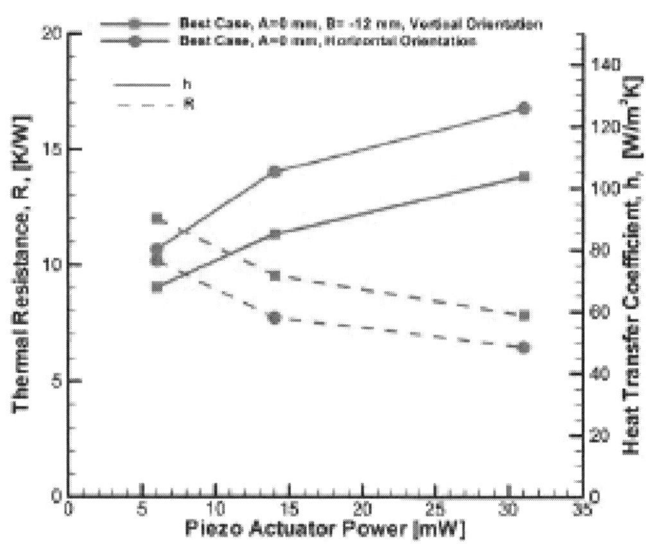

FIGURE 10.33 Thermal performance of best cases of horizontal and vertical orientations.

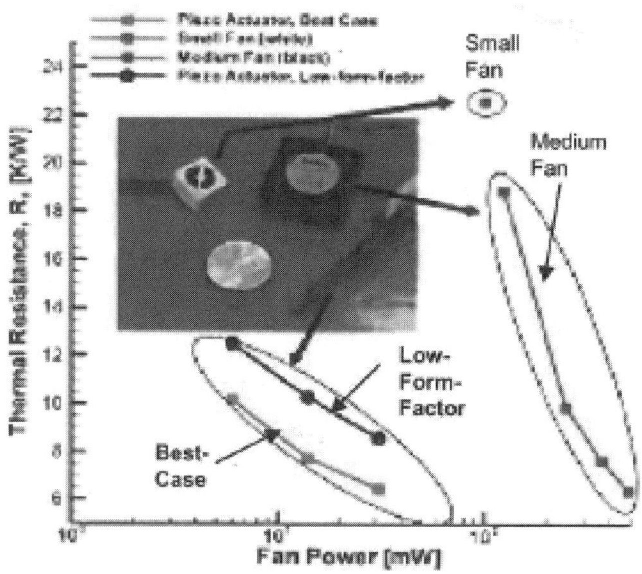

FIGURE 10.34 Thermal performance comparison between piezoelectric and small axial fans.

orientation is slightly better than that of the vertical orientation. For a piezoactuator power input of 31 mW, the calculated thermal resistances are 6.5°C/W and 7.8°C/W for the horizontal and vertical orientations, respectively. The comparison of the thermal performance of the best piezoelectric fan and two commercially available small axial fans is presented in Figure 10.34. As indicated in the figure, the piezoelectric fans not only have lower thermal resistance but also consume less power as compared to small axial fans. In summary, the piezoelectric fans have been employed as means to enhance local convective heat transfer while requiring only small amounts of power with lower acoustic noise.

ii. Electromagnetic actuator synthetic jets
Another way to create the synthetic jets is by a synthetic jet actuator. The synthetic jet actuator as shown in Figure 10.35 [28] consists of a cylindrical cavity with a sharp-edged orifice and a flexible diaphragm which is located on the opposite of the orifice. The bottom of the orifice extends outwards to provide a confined outflow condition. A synthetic jet comprises a train of vortices formed by successive ejection and suction of fluid across an orifice. The synthetic jet which is generated by pulse action of the diaphragm is impinging at the heated surface with the distance of H. When the jet is impinging onto a heated surface, the highly pulsatile nature promotes entrainment, mixing, and breaking up the thermal boundary layers that lead to a significant increase in the local heat transfer coefficient.

Figure 10.36 illustrates the flow pattern for a synthetic jet actuator operating in open ambient environment. The basic operation principle for

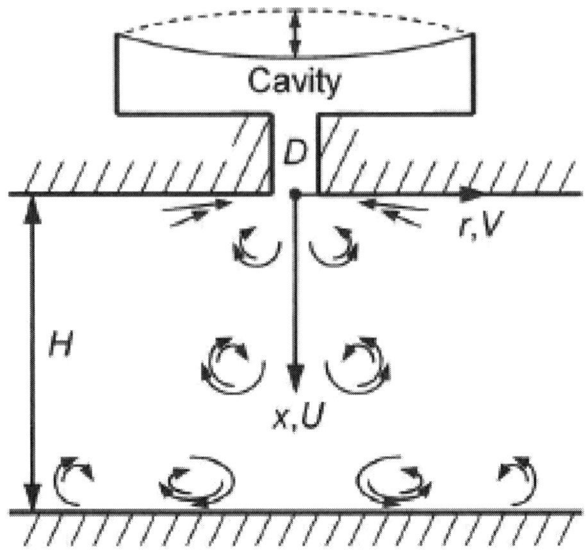

Figure 10.35 Schematic diagram of impinging synthetic jet.

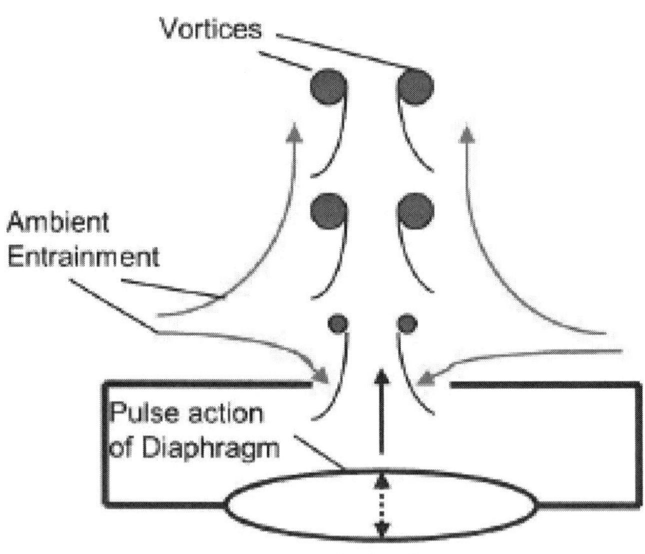

Figure 10.36 Synthetic jet actuator operation in open ambient environment.

a synthetic jet actuator is that a jet is generated due to the pulse action of the diaphragm. The surrounding air is sucked into the device from the orifice when the jet leaves the orifice with a train of vortices. The process is repeated again and again. The device itself creates no net mass flux.

The thermal performance of various cooling schemes, including a pair of synthetic jets and piezoelectric fan is given in Figure 10.36 [24]. As can be seen from the figure, the thermal performance of the synthetic jet is slightly better than that of the piezoelectric fan (Figure 10.37).

When a single impinging synthetic jet cools a heated surface as shown in Figure 10.35, a forced cross-flow is required to supply fresh cooling medium. However, Smith and Glezer [29] investigated the flow field of pair of adjacent synthetic jets. Persoons et al. [30] applies this vectoring effect in a pair of synthetic jets, thereby, combing (i) high local heat transfer rates and (ii) the cross-flow required for effective cooling. The purpose of their paper is to determine the optimal conditions for convective cooling as a function of jet-to-surface spacing and phase difference between the jets. Figure 10.38 shows the schematic of the test rig by Persoons et al. [30].

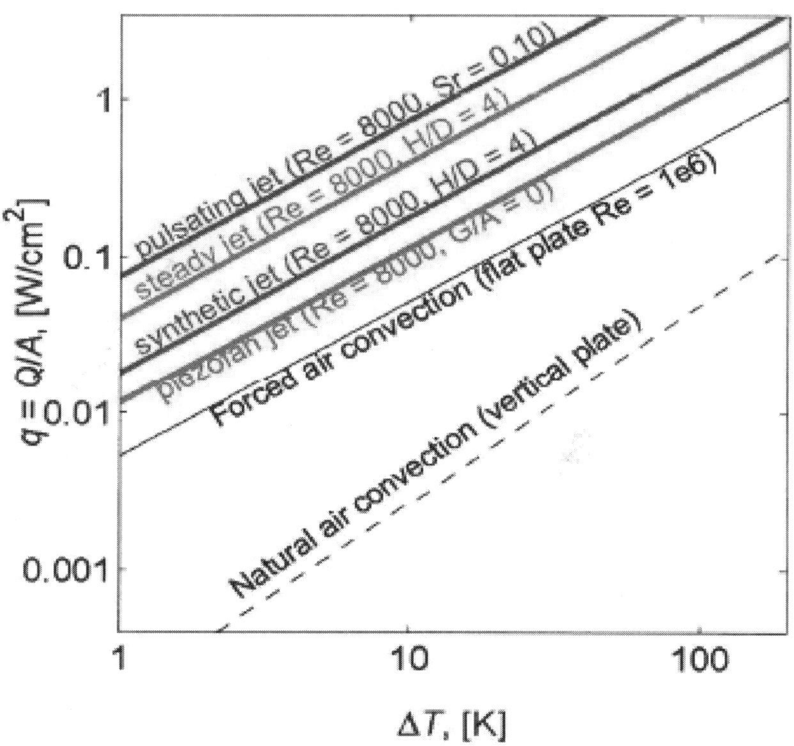

FIGURE 10.37 Thermal performance comparisons with synthetic jet.

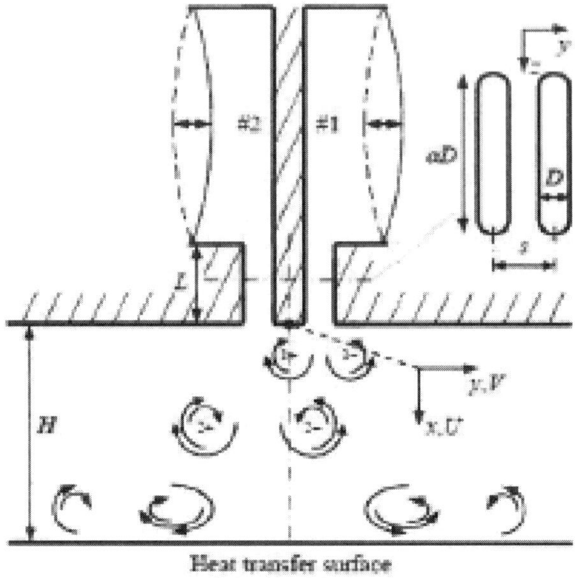

Figure 10.38 Schematic of test rig for a pair of synthetic jets.

In summary, both piezoelectric fan and synthetic jet actuator operate on the principle that jet generated periodically breaks up the thermal boundary layer on a heated surface and also induces significant entrainment on the surrounding fluid. However, the main difference is that piezoelectric fans are simpler, relying only on the vibrating motion of a cantilever beam to propel fluids towards a heated surface. One the other hand, a synthetic jet actuator generally requires three basic components: (1) an orifice to form vortices that make up the synthetic jet flow, (2) a cavity connected to the orifice, and (3) an oscillating diaphragm that drives the fluid motion. It should also be noted that both devices are commercially available.

4. Liquid cooling
   - Compact, cost-effective and reliable liquid cooling solutions
   - High performance and reliable pumps and condensers if needed
   - Development or improvement on thermal properties of new or existing dielectric fluids
     a. New technologies, including surface treatment to reduce temperature overshoot in pool boiling (direct immersion cooling) of dielectric fluids
   - Enhancement on microchannel cooling technologies to create low cost and reliable cold plates
     a. Mature microchannel cooling technology a must for 3D packaging
     b. Integrated microchannels cooled silicon chips
5. Other advanced cooling technologies
   - High performance, high capability micro heat pipes

  a. Embedded micro heat pipes for enhancing heat sink base plate spreading and also reducing base thickness
- High performance, high capacity long heat pipes or large vapor chambers
- High COP thermoelectric devices
  a. New thermoelectric materials and fabrication methods for performance improvement
  b. Thin film thermoelectric for TIM1 applications in device packages
- Technologies for refrigeration cooling
  a. Highly reliable miniaturized parts such as compressors, condensers and evaporators
  b. Low cost, low noise refrigerators using solid state, vapor compression, or absorption cycles

10.8 SUMMARY AND CONCLUSION

With increasing power on both component and system levels along with the demands of the reliability and performance, the thermal management becomes even more critical. The high airflow rates needed by the increased system power require the large high speed fans which not only consume more power but also increase acoustic noise. Power reduction strategies and high power thermal management become two of the critical elements for design of the future telecommunication equipment.

Air cooling has long been used for cooling of electronics, especially for the consumer electronics used in home and offices. It is simple, economical, and safe. Air cooling is also most commonly used for telecommunications equipment. While larger high speed fans may generate enough airflow to adequately cool the high power equipment, however, the air cooling is not recommended for any system over 20 KW because of the concerns about the fan power consumption which is directly related to the system COP and operation cost, and the excessive acoustic noise which often exceeds the required limits.

The operation cost of the data centers is skyrocketed in the recent years and it is also expected to be even higher in the future. The energy efficiency becomes the top priority for the data center operators and the equipment manufacturers. The methods to achieve the energy efficiency on the equipment have been discussed in the previous sections. Among them, the active liquid cooling is most attractive. The active liquid cooling not only can support even higher system power but also can eliminate the acoustic noise. And above all, the active liquid cooled rack is not only reliable but also energy efficient by eliminating all fans and the air filter which are required in the case of air cooling. It is believed that the active liquid cooling will be the choice of the future thermal management for the core telecommunication equipment. It is further expected the liquid cooled telecom equipment will be commercially available in the next few years.

While industry trends toward the high power systems, one should never ignore the value of the small box configuration equipment which can often be found in office, home and even in the corners of the hallways of the residential buildings.

From reliability and cost point views, passive thermal control should be adapted whenever possible. Efforts should also be considered to convert the fan-cooled small box configuration equipment to passive cooled system to achieve equipment energy efficiency for reduction in the operation cost and also to increase equipment reliability. The piezoelectric fans and synthetic jets discussed previously will have great benefits to low power and low profile boxes due to the enhancement on local heat transfer over hot spots.

Due to high-energy cost, alternative energies should be considered as the supplemental energy to reduce the operation cost of the equipment and the data centers. The thermoelectric power generation and the fuel cell technology appear to be feasible. Both of them provide clean energy with little or no pollution. Further researches to improve the thermal efficiency and the cost reduction will still be needed.

As discussed in previous sections, 1 W saved at the equipment will lead to a saving of more than 2.41 W in the data center. This clearly demonstrates that the energy efficiency at the system level has a great impact on the energy consumption and operation cost in the data center. With the cost of energy continuous increase, the energy efficiency becomes a must for the equipment manufacturers and the data center operators. Therefore, the energy efficiency should always be integrated into the system thermal designs both at the equipment level and the data center.

The first few papers in dealing with the thermal management and energy efficiency challenges in telecommunication systems and data centers was recently published by Garimella et al. [31, 32]. Though the paper is mainly focusing at the high-level discussions on the importance of the energy efficiency of telecommunication equipment and data centers; however, it does include a vast of data and information useful to the thermal engineers in these areas.

REFERENCES

1. Vukovic, A., "Power Density Challenges of Next Generation Telecommunication Networks", ElectronicsCooling, Vol. 9, No. 1, February, 2003.

1A. Gunther, Binns, Carmean, and Hall, "Managing the Impact of Increasing Microprocessor Power Consumption", Intel Technology Journal, Vol. 5, No. 1, February 2001.

2. "Datacom Equipment Power Trends and Cooling Applications", American Society of Heating, Refrigerating and Air-Conditioning Engineers, Inc, 2005.

3. Kenneth G. Brill, "2000–2010 Product Heat Density Trends Chart", White Paper, The Uptime Institute, 2006.

4. Telcordia Technologies Generic Requirements, GR-63-CORE, Issue 3, March, 2006.

5. Yeh, L. T.,"Thermal Management and Energy Efficiency of Telecommunication Equipment", Presented at 2nd Workshop on Thermal Management in Telecommunication Systems and Data Centers, Santa Clara, CA, April 25–26, 2012.

6. Zweben, C., "High Performance Thermal Management Materials", Advanced Packaging, February, 2006.

7. Tuckerman, D. B. and Pease, F. F., "High Performance Heat Sinking for VLSI", IEEE Electron Devices Letter, EDL-2, 1981.

8. Tuckerman, D. B., "Heat Transfer Microstructures for Integrated Circuits", Doctoral Thesis, Stanford University, 1984.
9. Alfieri, F., Tiwari, M. K., Zinovik, I., Poulikakos, D., Brunschwiler, T., and Michel, B., "3D Integrated Water Cooling of A Composite Multilayer Stack of Chips", Proceedings of 14th International Heat Transfer Conference, Washington, DC, USA, August 8–13, 2010.
10. Lee, T. and Lodhia, A., "Component and Board-Level Thermal Modeling Techniques for XFP Optical Transceivers", IMAPS Advanced Technology Workshop on Thermal Management for High Performance Computing and Telcom/Wireless Application, October 24–26, 2002.
11. Ellsworth, M. J., Jr., Campbell, L. A., Simons, R. E., Iyengar, M. K., Schmidt, R. R., Chu, R. C., The Evolution of Water Cooling for IBM Large Server Systems: Back to Future, Itherm, 2008.
12. Jei Wei, "Hybrid Cooling for Fujitsu Large Computer Systems", 2010 IEEE SPJW, Kyoto, Japan, 2010.
13. Escher, W., Wrunschwiler, T., Michel, B., and Poulikakos, D., "Experimental Investigation of an Ultra-thin Manifold Microchannel Heat Sink for Liquid-Cooled Chips", Intl. J. of Heat and Mass Transf., 2010 (current under review).
14. Escher, W., Michel, B., and Poulikakos, D., "Efficiency of Optimized Bifurcating Tree-like and Parellel Michrochannel Networks in the Cooling of Electronics", Intl. J. of Heat and Mass Transf. 52, 2009.
15. Bahadur, V., Hodes, M., Lyons, A., Krishnan, S., and Garimella, S. V., "Enhanced Cooling in a Sealed Cabinet using an Evaporating-Condensing Dielectric Mist", Proceedings of ITHERM, Orlando, FL, 2008.
16. Glover, G., "The Next Generation Router System Cooling", MS Thesis, California Polytechnic State University, 2009.
17. Schmidt, R. R., "Recent Advances and Future Challenges in Data Center", 7th International Business & Technology Summit, Natick, MA, August 22–23, 2007.
18. Brill, K. G., "Data Center Energy Efficiency and Productivity", White Paper, The Uptime Institute, 2007.
19. Graff, C., "Verizon's Thermal Management Program for Network Equipment Design", 2009.
20. Fisher, T., "Waste Heat Recovery from Electronics", CTRC Project review meeting, Purdue University, 2010.
21. Parsons, M. J. and Josefik, N. M., "Accelerating Production Readiness using Lean Product Development", Proceedings of ASME 2009 7th International Fuel Cell Science, Engineering and Technology Conference, Newport Beach, CA, June 8–10, 2009.
22. Kimber, M. L. and Garimella, S. V., "Measurement and Prediction of the Cooling Characteristics of a Generalized Vibrating Piezoelectric Fan", International Journal of Heat and Mass Transfer, Vol. 52, 4470–4478, 2009.
23. Kimber, M. L. and Garimella, S. V., "Cooling Performance of Arrays of Vibrating Cantilevers", Journal of Heat Transfer, Vol. 131, 111401, 2009.
24. Persoons, T., "Advanced Air Cooling Techniques for Thermal Management of Electronics, with a Focus on Synthetic Jets", 2nd Workshop on thermal Management in Telecommunication Systems and Data Center, Santa Clara, CA, April 25–26, 2012.

25. Acikalin, T., Wait, S. M., Garimella, S. V., and Raman, A., "Experimental Investigation of the Thermal Performance of Piezoelectrically Actuated Fans", Heat Transfer Engineering, Vol. 25, No. 1, 2004.
26. Acikalin, T. and Garimella, S. V., "Analysis and Prediction of the Thermal Performance of Piezoelectrically Actuated Fans", Heat Transfer Engineering, Vol. 30, No. 6, 2009.
27. Acikalin, T., Sauciuc, I., and Garimella, S. V., "Piezoelectric Actuators for Low-Form-Factor Electronics Cooling", IPACK2005-73288, 2005.
28. Persoons, T., McGuinn, A., and Murray, D. B., " A General Correlation for the Stagnation Point Nusselt Number of an Axisymmetric Impinging Synthetic Jet", International Journal of Heat and Mass Transfer, Vol. 54, 2011.
29. Smith, B. L., and Glezer, A., "Vectoring of Adjacent Synthetic Jets", AIAA Vol. 43 No. 10, 2005.
30. Persoons, T., O'Donvan, T. S., and Murray, D. B., "Heat Transfer in Adjacent Interacting Impinging Synthetic Jets", Paper No. HT2009-88440, ASME Summer Heat Transfer Conference, San Francisco, CA, July 19–23, 2009.
31. Garimella, S. V., Yeh, L. T., and Persoons, T., "Thermal Management Challenges in Telecommunication Systems and Data Centers", IEEE Transactions on Components, Packaging and Manufacturing Technology, Vol. 2, No. 8, 2012.
32. Garimella, S. V., Persoons, T., Weibe, J., and Yeh, L. T., "Technological Drivers in Data Centers and Telecom Systems: Multiscale Thermal, Electrical, and Energy Management", Applied Energy, Vol. 107, July 2013.

INDEX

Page numbers followed by f and t indicate figures and tables, respectively.

I

K

L

M

N

O

ABOUT THE AUTHORS

Lian-Tuu Yeh, Ph.D. and PE
ASME Fellow

Lian-Tuu Yeh received his BS in mechanical engineering from National Cheng Kung University, Taiwan, in 1967 and his M.S. and Ph. D in mechanical engineering from the University of Akron, Akron, Ohio in 1972 and 1976, respectively. Dr. Yeh has been practicing in thermal sciences for more than 35 years over diverse industries that include major companies such as Babcock & Wilcox, Texas Instruments, Lockheed-Martin, Boeing, Chiaro Networks, Fujitsu Networks and Futurewei Technologies in energy/power, electronics, aerospace, space and telecommunications industries. His work has been concentrated in thermal management of wide range of electronic systems. Dr. Yeh has made steady and significant contributions in the development of advanced cooling technologies for high-power electronic systems. His experience includes various types of liquid- or air-cooled electronic equipment, and above all, he has more than 10 years direct experience in thermal management of telecommunication systems. Dr. Yeh has been a member of Thermal Technology working group of INEMI (International/National Electronics Manufacturing Initiate) since its inception in 2002 and he has been serving as section lead for NETCOM since 2008. As an adjunct professor, Dr. Yeh taught the course of thermal analysis and design of electronic equipment for 2 years (1994, 1995) at the Department of Aerospace and Mechanical Engineering at the University of Texas at Arlington.

Dr. Yeh has been an active participant in and session, as well as conference organizer, of national and international heat transfer conferences. He has served as a technical reviewer for various heat transfer journals and conferences, and also for US National Science Foundation in reviewing proposals for research grants.

Dr. Yeh is a practicing engineer and is also researcher. He has published more than 62 technical papers in the field of heat transfer and a textbook entitled "Thermal Management of Microelectronic Equipment" by ASME Press in 2003. Dr. Yeh is also a co-author for the book chapter, "Thermal Management" of "flip chip advanced technologies" edited by Professor C.P. Wong and published by Springer Publishing in 2013. In addition, he also received one US patent for liquid cooling of electronic equipment.

Dr. Yeh was elected to a Fellow of the American Society of Mechanical Engineering (ASME) in 1990. He has served and is still an active member of the ASME K-16 Committee on Heat Transfer in Electronic Equipment and the former member of American Institute of Aeronautics and Astronautics (AIAA) Thermophysics Technical Committee. He has been listed in Who's Who in the South and Southwest (1988–1989) and Who's Who in the world (1994–1995). Dr. Yeh is also a registered professional engineer in the state of Texas.

Late Dr. Richard C. Chu
IBM Fellow Emeritus
Member, National Academy of Engineering, USA
Academician, Academia Sinica, ROC

Dr. Richard C. Chu is an IBM Fellow Emeritus in the Systems and Technology Group in Poughkeepsie, NY. He is an internationally recognized authority in cooling technology for high-performance computers. For more than four decades, he has played a key role in developing the world's most advanced computer cooling solutions.

Joining IBM in 1960 as a thermal engineer, he worked on the development of the IBM System/360. His 1965 invention of a multi-level air–liquid hybrid cooling design was pivotal to IBM's successful introduction of its System/360 Model 91. The introduction of this system marked the beginning of the water-cooling era which lasted throughout the years of bipolar technology.

Throughout his career, he has been a prolific technical innovator receiving 60 IBM invention achievement awards for over 200 patents and 150 invention publications. As one of the inventors of the cooling scheme for the IBM Thermal Conduction Module, he received an IBM Outstanding Innovation Award and a Corporate Award. This scheme and his modular cold plate cooling system concept formed the basis for the cooling design of IBM's largest computers for over 15 years. Variations of this cooling concept were adapted by other major mainframe computer manufacturers during this time period.

The importance of his many contributions to IBM and the industry led to his appointment as an IBM Fellow in 1983, the company's highest technical honor. He still continues in this capacity today, exploring cooling technology for future computers. In 1996, he initiated an effort to study the feasibility of using lower temperature cooling to enhance CMOS performance. He co-invented a dual-channel evaporator cold-plate enabling modular refrigeration cooling (MRC), which has been used on most IBM's CMOS-based high performance computers since 1997.

The importance of his contributions extends well beyond IBM. He has been a strong promoter of a close working relationship between industry and academia. For 25 years, he was the principal individual responsible for IBM's sponsorship of electronic cooling research at 12 universities. He is widely known for his active participation in ASME and IEEE conferences, presenting papers and often serving as a keynote speaker. He is a co-author of two books on electronic cooling. He also chaired a National Electronic Manufacturing Initiative (NEMI) technical working group in 2000 and 2002 to produce a Thermal Management Roadmap chapter for the NEMI Technology Roadmap.

He has been widely recognized for his contributions, receiving the ASME Heat Transfer Memorial Award, ITherm Memorial Award, SEMI-THERM Significant Contributor Award, InterPACK Conference Achievement Award, Chinese American Academic and Professional Society Achievement Award and the Chinese American Corporate Achievement Award from the Organization of Chinese Americans. In addition to being a member and past president of the IBM Academy of Technology, he is an ASME Fellow and a Fellow of the American Association for the Advancement of Science. He is the recipient of the Distinguished Alumni Award from both of his alma maters, the National Cheng-Kung University and Purdue University. He has also received honorary doctorate degrees from Purdue University and the American University of the Caribbean. Most recently he was honored as an Asian American Engineer of the year.